MASTERING ROBOTICS RESEARCH
FROM ENTHUSIAST TO EXPERT

4 BOOKS IN 1

BOOK 1
INTRODUCTION TO ROBOTICS RESEARCH: A BEGINNER'S GUIDE

BOOK 2
FUNDAMENTALS OF ROBOTICS RESEARCH: BUILDING A STRONG FOUNDATION

BOOK 3
ADVANCED TECHNIQUES IN ROBOTICS RESEARCH: BECOMING A SPECIALIST

BOOK 4
MASTERING ROBOTICS RESEARCH: FROM ENTHUSIAST TO EXPERT

ROB BOTWRIGHT

Published by Rob Botwright
Library of Congress Cataloging-in-Publication Data
ISBN 978-1-83938-513-1
Cover design by Rizzo

Disclaimer

The contents of this book are based on extensive research and the best available historical sources. However, the author and publisher make no claims, promises, or guarantees about the accuracy, completeness, or adequacy of the information contained herein. The information in this book is provided on an "as is" basis, and the author and publisher disclaim any and all liability for any errors, omissions, or inaccuracies in the information or for any actions taken in reliance on such information.

The opinions and views expressed in this book are those of the author and do not necessarily reflect the official policy or position of any organization or individual mentioned in this book. Any reference to specific people, places, or events is intended only to provide historical context and is not intended to defame or malign any group, individual, or entity.

The information in this book is intended for educational and entertainment purposes only. It is not intended to be a substitute for professional advice or judgment. Readers are encouraged to conduct their own research and to seek professional advice where appropriate.

Every effort has been made to obtain necessary permissions and acknowledgments for all images and other copyrighted material used in this book. Any errors or omissions in this regard are unintentional, and the author and publisher will correct them in future editions.

TABLE OF CONTENTS – BOOK 1 - INTRODUCTION TO ROBOTICS RESEARCH: A BEGINNER'S GUIDE

TABLE OF CONTENTS – BOOK 2 - FUNDAMENTALS OF ROBOTICS RESEARCH: BUILDING A STRONG FOUNDATION

TABLE OF CONTENTS – BOOK 3 - ADVANCED TECHNIQUES IN ROBOTICS RESEARCH: BECOMING A SPECIALIST

TABLE OF CONTENTS – BOOK 4 - MASTERING ROBOTICS RESEARCH: FROM ENTHUSIAST TO EXPERT

Introduction

Welcome to "Mastering Robotics Research: From Enthusiast to Expert," an extraordinary book bundle designed to be your guiding light through the fascinating and ever-evolving world of robotics. Whether you're an absolute novice taking your first steps or a seasoned enthusiast aiming to ascend to the highest echelons of expertise, this comprehensive collection of books has been meticulously crafted to meet you where you are and propel you toward mastery.

BOOK 1 - Introduction to Robotics Research: A Beginner's Guide
Our journey commences with Book 1, "Introduction to Robotics Research: A Beginner's Guide." Here, we will embark on an exploration of the fundamental principles that underpin the captivating field of robotics. This volume is the ideal starting point for those who are new to robotics research or wish to refresh their understanding of its core concepts. From the intriguing history of robotics to the essential terminologies and foundational knowledge, this book provides the sturdy stepping stones that will set the stage for your transformation into an adept researcher.

BOOK 2 - Fundamentals of Robotics Research: Building a Strong Foundation
In Book 2, "Fundamentals of Robotics Research: Building a Strong Foundation," we delve deeper into the mechanics of robotics. With a solid understanding of the basics in place, this volume delves into the intricate world of kinematics, dynamics, sensors, and actuators. As we explore these critical components, you'll gain the necessary theoretical groundwork that will empower you to navigate the complexities of robotic systems with confidence and clarity.

BOOK 3 - Advanced Techniques in Robotics Research: Becoming a Specialist
As you progress on your journey, Book 3, "Advanced Techniques in Robotics Research: Becoming a Specialist," beckons. Here, we venture into the realm of cutting-edge technologies and specialized techniques that define the modern landscape of robotics research. From computer vision and machine learning to advanced control systems, this volume offers an opportunity to deepen your expertise, setting you on the path to becoming a specialist in your chosen niche within robotics.

BOOK 4 - Mastering Robotics Research: From Enthusiast to Expert

The grand culmination of your odyssey awaits in Book 4, "Mastering Robotics Research: From Enthusiast to Expert." In this final volume, you will ascend to the zenith of your capabilities as a robotics researcher. Drawing upon the knowledge and skills acquired in the preceding books, we will explore complex research areas, tackle real-world challenges, and encourage you to innovate and create. By the end of this volume, you will have undergone a remarkable transformation, emerging as a confident and knowledgeable expert in the field of robotics research.

Each book in this bundle is designed to provide a rich blend of theoretical knowledge and practical application. We offer real-world examples, case studies, and hands-on projects to ensure that your learning journey is not confined to theory but extends into the realm of practical implementation.

Whether you dream of contributing groundbreaking research, pioneering technological advancements, or simply nurturing your boundless curiosity in the captivating world of robotics, this book bundle is your compass and your companion. Your journey from an enthusiast to an expert has just begun, and the horizons of robotics research are boundless.

As you dive into the pages of "Mastering Robotics Research: From Enthusiast to Expert," we encourage you to embrace the challenges, seize the opportunities, and allow your passion for robotics to propel you toward becoming a true master in this extraordinary field. Your adventure commences now.

BOOK 1
INTRODUCTION TO ROBOTICS RESEARCH
A BEGINNER'S GUIDE

ROB BOTWRIGHT

Chapter 1: The World of Robotics Research

In the realm of technology and innovation, few fields have captured the human imagination quite like robotics. From ancient automatons that sparked awe in the hearts of civilizations to the advanced humanoid robots that walk among us today, the history of robotics is a testament to humanity's unrelenting quest for innovation and progress.

Imagine a world where machines come to life, mimicking human and animal movements with uncanny precision. These early mechanical wonders, known as automata, date back to ancient civilizations in Egypt, Greece, and China. While they served various purposes, from entertainment to religious rituals, they all shared a common thread—capturing the essence of life through mechanical means.

As we embark on our journey through the historical milestones of robotics, we will encounter a tapestry of ingenuity, determination, and visionary thinking. From the first inklings of automation during the Industrial Revolution to the contemporary world of artificial intelligence (AI) and sophisticated robotic companions, each step in this journey has pushed the boundaries of what machines can achieve.

Our exploration begins with a deep dive into the early chapters of robotics, where the seeds of innovation were sown in ancient times. We'll witness the birth of automation during the Industrial Revolution, a period that reshaped industries and laid the foundation for programmable machinery.

But it doesn't stop there. The journey takes us through pivotal moments in time, such as the birth of the first industrial robot, the pioneering days of autonomous robotics, and the thrilling developments in space exploration. We'll delve into the world of cutting-edge humanoid robots and the integration of artificial intelligence, which has transformed the way robots perceive and interact with their surroundings.

We'll also explore how robots have infiltrated various sectors, from healthcare to consumer products, making our lives more convenient, safer, and efficient. In each chapter of our exploration, we'll uncover the remarkable stories behind these milestones, showcasing how robots have evolved from mere mechanical curiosities to indispensable companions in our daily lives.

As we embark on this journey, you'll gain a deeper appreciation for the innovators and visionaries who have shaped the world of robotics. From ancient inventors and engineers to contemporary pioneers, these individuals have left an indelible mark on our technological landscape. Their stories will inspire you, the budding robotics enthusiast, to become part of the next chapter in the history of robotics.

So, fasten your seatbelt, dear reader, as we embark on a captivating voyage through the annals of robotics. Let's explore the ingenious, the audacious, and the remarkable milestones that have defined this field and continue to drive it forward. Welcome to the journey of a robotics enthusiast—a journey through time and technology.

Early Automata: Think of ancient Egypt and Greece, where inventors created marvelous mechanical devices. These devices, known as automata, were early precursors to robots. They were designed to replicate human or animal movements, and they amazed people with their lifelike motions. These automata weren't just for entertainment; some were used in religious ceremonies and temples, showing that the fascination with robotics has ancient roots.

Industrial Revolution and Jacquard Loom: The Industrial Revolution brought about a revolution in machinery and automation. One standout innovation was the Jacquard loom, which used punched cards to control the weaving of intricate patterns. Although it might seem unrelated to modern robots, this concept of using instructions (in this case, punch cards) to control machinery was a precursor to modern programmable robots.

Unimate – The First Industrial Robot: In the early 1960s, George Devol and Joseph Engelberger introduced the Unimate, often considered the first industrial robot. This robotic arm was a game-changer for manufacturing. It could perform tasks like loading and unloading heavy parts in factories. It not only increased efficiency but also made workplaces safer by taking on dangerous tasks.

Shakey the Robot: Imagine Shakey as a pioneer in the world of autonomous robots. Developed in the late 1960s and early 1970s, Shakey was a mobile robot that could reason about its actions and navigate through its environment. It could solve problems and plan its movements. This was a crucial step toward robots becoming more autonomous and capable of complex tasks.

The Robot Arm on the Space Shuttle: Picture the iconic Canadarm, a robotic arm used on the Space Shuttle. Developed in 1981, it was like a giant, remote-controlled crane in space. It played a crucial role in capturing and deploying satellites, aiding astronauts during spacewalks, and even helping assemble the International Space Station (ISS). It showcased how robots could excel in space exploration.

The Mars Rovers: Think of NASA's Mars rovers, which include names like Sojourner, Spirit, Opportunity, and Curiosity. These robots have explored the Martian surface, conducting experiments and sending back valuable data. They've demonstrated the capability of remote-controlled robots in navigating harsh and distant environments.

Boston Dynamics and Advanced Robotics: Let's fast forward to more recent times. Boston Dynamics has been making waves in robotics with robots like BigDog, Atlas, and Spot. These robots showcase extraordinary mobility and agility. They can perform tasks that were once thought impossible for robots, such as running, jumping, and even dancing. Their development is a testament to how robots are becoming more versatile and capable.

Soft Robotics: Think about robots that are soft, flexible, and adaptable, inspired by nature. These robots are made from materials that can deform and bend. They are perfect for applications like medical robotics, where delicate movements are essential. Imagine a soft robotic arm gently assisting in surgeries, offering a new level of precision.

Autonomous Vehicles: While not often thought of as robots, self-driving cars and autonomous vehicles are indeed a form of robotics. Companies like Tesla and Waymo are pushing the boundaries of technology to create vehicles that can drive themselves. This innovation could revolutionize transportation, making it safer and more efficient.

Humanoid Robots: Envision humanoid robots like ASIMO and Sophia. They are designed to look and sometimes even behave like humans. These robots represent advancements in natural language processing, computer vision, and social interaction. Imagine having a conversation with a robot like Sophia—it's like a glimpse into the future of AI and robotics.

Robotic Surgery: Think about how robots are transforming surgery. The da Vinci Surgical System, for instance, allows surgeons to perform complex procedures with incredible precision and minimal invasiveness. It's like having a robotic assistant in the operating room, enhancing the capabilities of human surgeons.

Robots in Space: Consider the robots that explore outer space. They've become indispensable for missions to celestial bodies. From the Mars rovers to the Mars Helicopter (Ingenuity), which recently achieved powered flight on Mars, these robots are trailblazers in interplanetary exploration.

AI Integration: Think about how artificial intelligence (AI) has become an integral part of robotics. Machine learning and deep learning algorithms enable robots to perceive and adapt to their surroundings. They can make decisions based on what they see and learn from their experiences. It's like giving robots a form of intelligence.

Robots in Healthcare: Imagine robots assisting in healthcare. They can provide patient care, help with rehabilitation, deliver medication, and even assist in surgeries. It's like having a helpful assistant in a hospital, making healthcare more efficient and precise.

Consumer Robots: Finally, consider the robots that have become part of our everyday lives. Robot vacuum cleaners like Roomba, personal assistants like Amazon's Alexa, and even robotic pets have found their way into our homes. They make our lives more convenient and, in some cases, offer companionship.

These historical milestones are not just about technology; they represent the human imagination, creativity, and determination to push the boundaries of what robots can do. Robotics is a field with a rich history and an exciting future, and these milestones are markers along the way.

Robotics, is a captivating field that brings together a diverse array of knowledge and expertise. It's like a grand orchestra, where different instruments, each representing a unique discipline, play in harmony to create something extraordinary— robots. So, let's explore why robotics is so multidisciplinary and how this multidimensionality enriches its innovations.

Engineering and Mechanical Design: Picture engineers meticulously sketching and designing the physical structure of a robot. They consider factors like materials, mechanics, and durability. The robot's frame, joints, and actuators are all products of engineering marvels. It's akin to crafting the skeleton and muscles of a living being.

Electronics and Electrical Engineering: Now, think of the electrical engineers who infuse life into robots through circuits and sensors. They design the electronic nervous system that enables robots to sense and respond to their surroundings.

From microcontrollers to sensors, it's the electrical magic that makes robots perceptive.

Computer Science and Software Engineering: Next, imagine the realm of computer science. Here, software engineers write lines of code that govern a robot's behavior. It's like the brain of the robot, instructing it on how to move, react, and make decisions. Algorithms are the language through which robots comprehend the world.

Mathematics: Behind the scenes, mathematics plays a significant role. Mathematicians help solve complex equations for robot kinematics, dynamics, and control. They ensure that a robot's movements are precise, its balance is maintained, and its actions are optimal. It's like the mathematical choreography of a robot's dance.

Physics: Physics, the science of motion and forces, is integral too. Physicists help determine how a robot interacts with its environment. They calculate the impact of gravity, friction, and other physical factors on a robot's movement. It's akin to predicting the path of a projectile.

Biology and Biomimicry: Here's a twist—biological knowledge also seeps into robotics. Biomimicry is the art of imitating nature. Scientists and engineers often draw inspiration from living creatures to create robots with lifelike features. Think of robots that mimic the agility of animals or the dexterity of human hands.

Material Science: Material scientists are like wizards who conjure up innovative materials that robots are made of. From super-strong alloys to flexible polymers, these materials ensure robots can withstand harsh conditions or perform delicate tasks.

Aerospace Engineering: In the world of drones and flying robots, aerospace engineering takes center stage. Engineers in this field design robots that soar through the skies, exploring remote locations and capturing breathtaking aerial views.

Ethics and Philosophy: Here's something intriguing—ethics and philosophy play a part too. Thinkers and ethicists ponder questions like the moral responsibility of robots or the implications of AI. It's like asking if robots have a sense of right and wrong.

Environmental Science: Robots are also champions of environmental science. They're used to monitor and protect fragile ecosystems, analyze water quality, or even clean up pollution. These robots are like environmental detectives safeguarding our planet.

Medical Science: In the realm of medical robotics, doctors, and medical professionals team up with engineers to create surgical robots that enhance precision in the operating room. These robots assist in delicate procedures, making surgeries less invasive and more effective.

Psychology and Human-Robot Interaction: Ever heard of human-robot interaction? Psychologists and experts in this field study how humans and robots can collaborate effectively. They delve into the psychology of trust, communication, and collaboration between humans and machines.

Artificial Intelligence (AI): Ah, we can't forget AI! AI researchers work on endowing robots with the ability to learn, adapt, and make decisions independently. This is where machine learning, neural networks, and deep learning come into play. Robots are becoming smart, learning from experience, just like humans.

Economics: Economics even plays a role in robotics. Economists study the impact of automation on the job market, the cost-effectiveness of robots in various industries, and the potential for economic growth through automation.

Art and Design: Artists and designers work their magic too. They make robots aesthetically pleasing or design interactive installations that blend technology and art. It's like turning robots into works of art that engage our senses and emotions.

Space Exploration: In the world of space exploration, astronomers, physicists, and aerospace engineers collaborate to design robots for planetary exploration. These robots venture into the unknown, collecting data from distant celestial bodies.

So, you see, robotics isn't confined to a single discipline. It's an intricate tapestry of knowledge, where experts from various fields collaborate to create robots that revolutionize our world. From the smallest nanobots to towering humanoid robots, each is a testament to the power of multidisciplinary teamwork and innovation.

In this book, we'll embark on a journey through the multidisciplinary landscape of robotics. We'll explore how each discipline contributes to the creation of robots that can explore the depths of the ocean, perform intricate surgeries, assist in disaster recovery, or even become our companions in everyday life.

As we delve deeper into this multidisciplinary realm, you'll gain a profound appreciation for the synergy of knowledge and expertise that brings robots to life. So, fasten your seatbelt and get ready to explore the diverse and exciting world of robotics, where collaboration and innovation know no bounds.

As we dive into the world of robotics today, you'll be delighted to know that there are some remarkable trends shaping the field.

1. Automation and Industry 4.0: Imagine factories where robots work hand-in-hand with humans. This trend, known as Industry 4.0, is revolutionizing manufacturing. Robots equipped with advanced sensors and AI are optimizing production lines, increasing efficiency, and ensuring product quality.

2. AI-Powered Robots: Think of robots becoming smarter than ever, thanks to artificial intelligence. Machine learning and deep learning algorithms enable robots to learn from data, adapt to changing environments, and make decisions on their own. It's like giving robots a dose of human-like intelligence.

3. Collaborative Robots (Cobots): Cobots are robots designed to work alongside humans safely. Picture a factory worker and a robot collaborating on intricate tasks. This trend fosters human-robot teamwork, making manufacturing more flexible and safer.

4. Autonomous Vehicles: Autonomous cars are no longer science fiction. Companies like Tesla and Waymo are developing self-driving vehicles that can navigate complex traffic scenarios. It's like having a chauffeur robot take you to your destination.

5. Service Robots: Imagine robots in various service roles. There are robots that assist in healthcare, deliver packages, serve in restaurants, and even clean our homes. These robots are like helpful companions that make our lives easier.

6. Robotics in Space: Robots have also conquered space exploration. Think of Mars rovers like Curiosity and Perseverance, which are helping us understand the Red Planet. We're even planning missions to send robots to the Moon and beyond.

7. Soft Robotics: Soft robots are a fascinating trend. These robots are made from flexible materials, mimicking the softness and adaptability of living organisms. They're perfect for tasks like delicate surgeries and exploring fragile environments.

8. Swarm Robotics: Imagine a swarm of tiny robots working together, much like a colony of ants. Swarm robotics is all about coordinated teamwork among many small robots. This trend has applications in search and rescue, agriculture, and even environmental monitoring.

9. Robot Learning from Human Demonstrations: Robots can now learn by watching humans. For example, a robot can observe a human assembling a product and replicate the task.

It's like teaching robots through example, making them more versatile.

10. Exoskeletons: Exoskeletons are wearable robots that can enhance human strength and mobility. They're like a suit of armor that can assist people with physical disabilities or help workers lift heavy objects.

These trends are not just shaping the present but also paving the way for a future where robots become integral to our daily lives. They represent the fusion of technology, innovation, and creativity.

Challenges in Robotics

But, of course, along with these exciting trends come challenges. It's like a thrilling adventure with a few obstacles along the way.

1. Safety: Safety is paramount. As robots become more autonomous and work closely with humans, ensuring their safety and preventing accidents is a significant challenge. It's like making sure our robotic companions don't inadvertently harm us.

2. Ethical Dilemmas: Imagine the ethical questions that arise. For instance, should robots have rights? How do we address concerns about privacy when robots become more integrated into our lives? These are complex moral dilemmas we must grapple with.

3. Job Displacement: The fear of robots taking over jobs is a concern. While automation can boost productivity, it can also lead to job displacement in certain industries. It's like finding a balance between efficiency and employment.

4. Cost and Accessibility: Developing and deploying advanced robots can be costly. Ensuring that these technologies are accessible and affordable to a wide range of industries and individuals is a challenge.

5. Human-Robot Interaction: Creating seamless communication between humans and robots is vital. We need to ensure that robots understand human commands and intentions accurately. It's like teaching robots to speak our language, both figuratively and literally.

6. Data Privacy: As robots collect and process vast amounts of data, concerns about data privacy and security come to the forefront. It's like safeguarding our digital footprint in an increasingly connected world.

7. Technical Challenges: Building robots that can operate in diverse environments, from outer space to underwater, presents technical hurdles. It's like conquering the vastness of nature with technology.

8. Environmental Impact: We also need to consider the environmental impact of robotics. From energy consumption to the disposal of robotic components, sustainability is a concern.

9. Regulation and Standards: Creating regulations and standards for robotics is an evolving challenge. Ensuring that robots adhere to ethical and safety guidelines is crucial.

10. Cultural Acceptance: Cultural attitudes toward robots can vary widely. Some cultures embrace robotic technology, while others may have reservations. It's like navigating cultural diversity in the world of technology.

In essence, robotics is a field brimming with potential and promise. The trends we're witnessing today are reshaping industries, advancing science, and enriching our lives in countless ways. Yet, it's important to recognize and address the challenges that come with these innovations.

As we continue our journey through the world of robotics in this book, we'll explore how these trends and challenges shape our future. We'll delve deeper into each trend, unraveling the intricacies and marveling at the possibilities. So, fasten your seatbelt for an exciting ride through the world of robots, where trends meet challenges, and innovation knows no bounds.

Okay, let's break down robots into their fundamental components and systems. Robots are like intricate machines with different parts working in harmony to achieve various tasks. Here's a glimpse into the inner workings of these mechanical marvels.

1. Mechanical Structure: Imagine the skeleton of a robot; that's its mechanical structure. It's like a framework that provides support and shape. Robots can have various forms, from industrial arms with multiple joints to humanoid robots with limbs resembling ours.

2. Actuators: Actuators are like the muscles of a robot. They make the robot move. Imagine a robotic arm reaching out to pick up an object. Electric motors, pneumatic cylinders, and hydraulic systems are common actuators that generate motion in different types of robots.

3. Sensors: Sensors are the robot's senses. They perceive the world around them. Picture a robot using sensors to detect obstacles, measure distances, or identify objects. Sensors include cameras, ultrasonic sensors, infrared sensors, and more.

4. Controllers: Controllers are like the brains of the operation. They process information from sensors and decide how the robot should respond. It's like the robot's decision-making center. Controllers can be microcontrollers, microprocessors, or even full-fledged computers.

5. Power Source: Robots need energy to operate. Think of batteries, power cables, or even fuel cells as the robot's energy source. These power sources provide the necessary electricity, pneumatic pressure, or hydraulic fluid to drive the actuators and electronics.

6. End Effectors: End effectors are like a robot's specialized hands or tools. They're what the robot uses to interact with its environment. Imagine a robotic gripper picking up objects or a welding tool used in industrial robots. End effectors can be anything from claws to lasers.

7. Communication Systems: Robots often need to communicate with humans or other robots. Think of Wi-Fi or

Bluetooth connections that allow robots to send and receive data. It's like a robot's way of talking to the world.

8. Control Software: Control software is the set of instructions that tell the robot what to do. It's like the robot's choreography. Engineers write programs that guide the robot's movements, reactions, and decision-making.

9. Feedback Systems: Robots need to know if they're doing things correctly. Feedback systems, like encoders or sensors, help the robot understand its position, speed, and whether it's achieved its goals. It's like a robot's way of self-assessment.

10. Navigation Systems: For robots that move around, navigation is crucial. Imagine a robot vacuum cleaner mapping your home or a self-driving car navigating city streets. Navigation systems can include GPS, lidar, or cameras for vision-based navigation.

11. Human-Machine Interface: If a robot interacts with humans, it needs a way to understand and respond to human input. Touchscreens, voice recognition, or gesture control can serve as interfaces. It's like speaking a common language with the robot.

12. Safety Systems: Safety is paramount. Robots need systems to detect potential hazards and ensure they operate safely. Emergency stop buttons, protective barriers, and collision detection sensors are examples of safety measures.

13. Mobility Systems: Some robots need to move from one place to another. Wheeled robots use wheels, while legged robots use legs for mobility. It's like how we use our legs to walk or wheels to drive.

14. Perceptual Systems: Perceptual systems enable robots to perceive the world through sensors like cameras, microphones, and touch sensors. These systems help robots understand their surroundings and react accordingly.

15. Feedback Control Loops: Imagine a thermostat maintaining a constant temperature in your home. Feedback control loops work similarly in robots. They continuously adjust the robot's actions based on feedback from sensors, ensuring precise and stable performance.

16. Kinematic Chains: For robots with multiple joints like robotic arms, think of kinematic chains as their "skeleton." These chains determine how the joints move relative to each other, allowing for precise control of the robot's movements.

17. Gripping and Manipulation Systems: Robots often need to pick up, manipulate, or assemble objects. Gripping and manipulation systems, like robotic hands or claws, enable them to interact with their environment.

18. Localization and Mapping: Robots that navigate complex environments need to know where they are and what's around them. Localization and mapping systems create a map of the robot's surroundings and help it determine its position within that map.

19. Swarm Control Systems: In the world of swarm robotics, robots work together like a swarm of bees. Swarm control systems coordinate the actions of multiple robots, enabling them to collaborate on tasks such as exploration or search and rescue.

20. Redundancy and Fault Tolerance: Robots are designed with redundancy and fault tolerance in mind to ensure that they can continue functioning even in the presence of component failures. Think of it as having backup

Chapter 2: Foundations of Robotics Technology

When it comes to robots, we're talking about the nuts and bolts, quite literally! Robots are sophisticated machines with a complex array of hardware and actuators that make them come to life.

Mechanical Structure: First and foremost, let's consider the skeleton of a robot. This is the mechanical structure that provides the robot with its shape and support. Think of it as the framework upon which all other components are mounted. Robots come in various forms, from industrial arms with multiple joints to wheeled rovers and humanoid robots with limbs resembling ours.

Actuators: Now, let's talk about what gives a robot its motion—actuators. These are the muscles of the robot, so to speak. Actuators generate physical movement, making the robot's various parts come to life. Whether it's the rotation of a robotic arm, the swaying of a drone's propellers, or the locomotion of a robot on wheels or legs, actuators are responsible for these motions. They can take many forms, including electric motors, pneumatic cylinders, and hydraulic systems, each suited to specific applications.

Sensors: Moving on, let's consider the robot's senses—the sensors. Sensors are what allow robots to perceive the world around them. Just like our eyes, ears, and skin help us understand our environment, sensors provide robots with data about theirs. These sensors can include cameras, ultrasonic sensors, infrared sensors, touch sensors, and more. They allow robots to see, hear, touch, and even smell their surroundings, enabling them to interact with the world in various ways.

Controllers: Now, imagine a robot's brain. This is where controllers come into play. Controllers are the decision-makers of the robot, much like our own brains. They process information received from sensors and determine how the robot should respond. Depending on the complexity of the robot, controllers can range from microcontrollers and microprocessors to advanced computing systems. They are responsible for orchestrating the robot's movements, actions, and decision-making processes.

Power Source: Every robot needs a source of energy to operate. Think of this as the robot's lifeblood. Robots can be powered by various means, including batteries, power cables, or even fuel cells. These power sources supply the necessary electricity, pneumatic pressure, or hydraulic fluid to drive the actuators and electronics, keeping the robot functional.

End Effectors: Robots often need specialized tools or hands to interact with their environment. These tools are known as end effectors. Picture a robotic gripper picking up objects, a welding tool in an industrial robot, or even a robotic hand designed for precise surgical procedures. End effectors are like the fingertips of a robot, allowing it to perform a wide range of tasks.

Communication Systems: In our interconnected world, communication is key. Robots are equipped with communication systems that enable them to exchange data with humans, other robots, or control systems. These systems can include Wi-Fi, Bluetooth, or even wired connections, depending on the robot's application. It's like the robot's way of talking to the world and receiving commands.

Control Software: Just as we have a set of instructions for various tasks, robots have control software that tells them what to do. Engineers write programs that guide the robot's movements, actions, and decision-making processes. It's like the choreography that governs a robot's performance, ensuring it carries out tasks with precision.

Feedback Systems: Robots need to know if they're doing things correctly. Feedback systems provide information about the robot's own state or performance. They help the robot understand its position, speed, and whether it has achieved its goals. Think of this as the robot's way of self-assessment, allowing it to make adjustments as needed.

Navigation Systems: For robots that move through space, navigation is essential. Whether it's a self-driving car on city streets or a drone exploring remote landscapes, navigation systems provide robots with the ability to determine their position and plan routes. These systems can include GPS, lidar, cameras for vision-based navigation, and more.

Human-Machine Interface: When robots interact with humans, they need a way to understand and respond to human input. This is where human-machine interfaces come into play. Imagine touchscreens, voice recognition systems, or gesture control mechanisms that allow humans to communicate with robots naturally. These interfaces facilitate seamless interactions between humans and machines.

Safety Systems: Safety is a top priority when it comes to robots. Safety systems are designed to detect potential hazards and ensure that robots operate without causing harm to humans or themselves. These systems can include emergency stop buttons, protective barriers, and collision detection sensors.

Environmental Adaptation: Depending on their intended tasks, robots may need to adapt to different environments. Think of underwater robots exploring the depths of the ocean or robots designed for space exploration. Robots built for specific environments require adaptations such as waterproofing, radiation shielding, or extreme temperature tolerance.

Redundancy and Fault Tolerance: Robots are often equipped with redundancy and fault-tolerant systems to ensure continued operation even in the presence of component failures. These systems are like backup mechanisms, ensuring that the robot remains functional and safe.

Mobility Systems: Some robots need to move from one place to another. Wheeled robots use wheels for mobility, while legged robots use legs. Mobility systems determine how the robot moves through its environment, allowing it to navigate terrain and reach its destination.

Perceptual Systems: Perceptual systems enable robots to perceive the world through their sensors. Just as we rely on our senses to understand our surroundings, robots rely on sensors like cameras, microphones, and touch sensors to gather information about their environment.

Feedback Control Loops: Feedback control loops are like an orchestra conductor ensuring that all the components of a robot work in harmony. These control loops continuously adjust the robot's actions based on feedback from sensors, ensuring that the robot performs its tasks with precision and stability.

Kinematic Chains: For robots with multiple joints, such as robotic arms, kinematic chains determine how the joints move relative to each other. These chains are essential for precise control of the robot's movements and are like the robot's internal skeletal structure.

Gripping and Manipulation Systems: Robots often need to manipulate objects, whether for assembly, pick-and-place tasks, or even delicate surgeries. Gripping and manipulation systems, such as robotic hands or claws, enable robots to interact with objects in their environment.

Localization and Mapping: For robots that navigate complex environments, localization and mapping systems are crucial. These systems create maps of the robot's surroundings and help it determine its precise position within that map. Think of it as a GPS system for robots.

Swarm Control Systems: In the fascinating field of swarm robotics, multiple robots work together in a coordinated manner, much like a swarm of bees. Swarm control systems coordinate the actions of these robots, allowing them to collaborate on tasks such as exploration or search and rescue.

Redundancy and Fault Tolerance: Robots are often equipped with redundancy and fault-tolerant systems to ensure continued operation even in the presence of component failures. These systems are like safety nets, providing backup mechanisms to keep the robot functional.

When we talk about robots, it's not just about their physical hardware; it's also about the software that drives them and the control systems that guide their actions. It's like the intricate dance choreography that brings a robot's movements to life.

Control Software: At the heart of every robot is its control software, the set of instructions that tell it what to do. Think of it as the robot's brain, responsible for coordinating its movements, actions, and decision-making processes. This control software is meticulously crafted by engineers and programmers to ensure that the robot operates precisely and efficiently.

Programming Languages: Just as humans communicate through languages, robots have their own programming languages. Engineers write code in languages like Python, C++, or Java to instruct the robot. These languages provide a structured way to convey the robot's tasks and behaviors. It's like giving the robot a language it understands.

Algorithmic Control: Robots rely on algorithms, step-by-step sequences of instructions, to perform specific tasks. Imagine a robot following an algorithm to navigate a maze or assemble parts on an assembly line. These algorithms are like the recipe that guides a robot through its tasks.

Motion Planning: Consider a robot that needs to move from point A to point B without colliding with obstacles. Motion planning algorithms help the robot calculate the safest and most efficient path to its destination. It's like a robot's GPS for navigating its environment.

Feedback Control: Feedback control is a crucial aspect of robot software. It involves continuously adjusting the robot's actions based on feedback from its sensors. Imagine a robot with a camera that uses feedback to adjust its aim and capture a target accurately. This feedback loop ensures that the robot's movements are precise and responsive.

Inverse Kinematics: For robots with multiple joints, like robotic arms, inverse kinematics is essential. It's the mathematical technique that calculates the joint angles required to achieve a specific end-effector position. Think of it as the robot's way of solving puzzles to position its arm precisely.

Machine Learning and AI: The field of robotics is increasingly embracing machine learning and artificial intelligence (AI). Machine learning algorithms enable robots to learn from data and adapt to changing environments. Consider a robot that learns to recognize objects through repeated exposure—this is the magic of AI in robotics.

Neural Networks: Neural networks, inspired by the human brain, are a subset of AI used in robotics. They enable robots to process complex data, such as images or natural language, to make decisions. Think of it as the robot's ability to think and learn like a human.

Sensor Fusion: Robots often have multiple sensors, each providing different types of information. Sensor fusion is the process of combining data from various sensors to create a more comprehensive understanding of the robot's surroundings. It's like the robot's way of seeing the bigger picture.

Localization and Mapping (SLAM): For robots that navigate complex environments, simultaneous localization and mapping (SLAM) is essential. This technique allows a robot to create a map of its surroundings while simultaneously determining its own position within that map. Imagine a robot exploring an unknown terrain—it uses SLAM to understand where it is and what's around it.

Behavior-Based Control: Some robots use behavior-based control systems. Instead of following explicit programming, they exhibit behaviors based on their sensors and environment.

It's like a robot making decisions on the fly, responding to what it encounters in real-time.

Human-Robot Interaction (HRI): Robots often need to interact with humans, and HRI is a critical aspect of their software. This involves designing interfaces and algorithms that enable natural and safe interactions between humans and robots. Think of voice assistants or robots that collaborate with factory workers—it's all about effective communication.

Safety Protocols: Ensuring the safety of robots and those around them is paramount. Software plays a vital role in implementing safety protocols. For example, emergency stop procedures, collision avoidance algorithms, and obstacle detection are all part of a robot's safety measures.

Real-Time Control: Many robots operate in real-time environments where split-second decisions are crucial. Real-time control systems ensure that the robot can respond to changing conditions instantaneously. It's like the reflexes of a robot, allowing it to react swiftly to unexpected situations.

Distributed Control: In scenarios where multiple robots work together, distributed control systems come into play. These systems coordinate the actions of multiple robots to achieve a common goal. Think of a team of drones performing a synchronized aerial display—distributed control ensures they move in harmony.

Environmental Adaptation: Robots may need to adapt to different environments. Software enables them to adjust their behavior and parameters based on their surroundings. It's like a robot switching to different modes when it moves from land to water or enters a hazardous area.

Path Planning and Optimization: Path planning software helps robots find the most efficient route to reach their destination while avoiding obstacles. Imagine a delivery robot mapping out the quickest path to deliver a package—it optimizes its journey for efficiency.

Simulation and Testing: Before deploying a physical robot, engineers often use software simulations to test and refine their algorithms. These virtual environments allow them to experiment with different scenarios and fine-tune the robot's behavior.

Cloud-Based Robotics: With the advent of cloud computing, robots can tap into the power of remote servers for data processing and complex computations. This enables robots to perform tasks that require vast computational resources, like natural language processing or image recognition.

Adaptive Control: Adaptive control systems allow robots to adjust their behavior based on changing conditions. Imagine a robot that learns to adapt to varying terrain while traversing uneven landscapes—it continually refines its movements to maintain stability.

Human-Centric Software: In applications where robots assist or collaborate with humans, the software is designed with a human-centric approach. It focuses on making the robot's behavior understandable and predictable to humans, ensuring safe and effective cooperation.

Data Management: Robots generate and collect vast amounts of data. Data management software handles the storage, processing, and analysis of this data, providing valuable insights into the robot's performance and its environment.

Autonomous Navigation: Autonomous robots rely heavily on software for navigation. They use algorithms and sensor data to navigate complex environments, avoiding obstacles and reaching their goals independently.

Software Updates and Maintenance: Just like our devices receive software updates, robots require periodic updates and maintenance to stay efficient and secure. These updates can introduce new features, fix bugs, or enhance the robot's capabilities.

Linear Algebra for Robotic Applications

Imagine, if you will, that we're embarking on an exciting adventure into the heart of mathematics—specifically, linear algebra—and its indispensable role in robotics.

Vectors and Scalars: In the realm of linear algebra, we encounter two fundamental entities: vectors and scalars. Scalars are like solo players, representing quantities with just a magnitude—think of temperature or speed. Vectors, on the other hand, are dynamic duos, as they not only have magnitude but also direction. Robots use vectors to describe everything from positions in space to velocities and forces.

Vector Operations: Picture vectors as arrows in space. Linear algebra equips us with operations like addition and subtraction for vectors. When robots move, their positions change, and these changes can be calculated using vector operations. Imagine a robot plotting its course by adding vectors that represent motion in different directions—this is where the magic begins.

Dot Product: The dot product is like a robotic handshake between two vectors. It allows us to find the angle between them and quantify how much they align. For robots, this is crucial in understanding how forces act on them or how much they move in a particular direction.

Cross Product: The cross product is another vector operation that robots find handy. It's like a robotic high-five. It yields a vector perpendicular to the plane formed by two input vectors, making it useful in applications like calculating torques or determining orientations.

Matrices: Now, let's venture into the world of matrices. Matrices are like grids filled with numbers, and they offer a compact way to represent data. Robots use matrices to store information about transformations, such as rotations and translations. Imagine a matrix as a set of instructions that tells a robot how to move or change its orientation.

Matrix Multiplication: Matrix multiplication is like the intricate dance of robots collaborating to perform a task. It combines matrices to represent complex transformations. For example, imagine a robot picking up an object and rotating it while moving to a new location—matrix multiplication orchestrates this dance of motions.

Determinants: Determinants are mathematical determiners of matrices' properties. They're like quality control inspectors for robot instructions. Robots use determinants to check if a set of transformations is reversible or to ensure that their actions won't lead to unexpected consequences.

Eigenvalues and Eigenvectors: Eigenvalues and eigenvectors are like the secret sauce in robotics. They help robots understand stable states and directions in complex systems. For instance, imagine a robot trying to balance on one wheel— it relies on eigenvalues and eigenvectors to maintain stability.

Linear Independence: Linear independence is the notion of vectors not being redundant. Think of it as a robot choosing the most efficient set of instructions to achieve a goal. Robots use linear independence to avoid unnecessary movements and optimize their actions.

Span and Basis: Span is like the playground where robots explore their possibilities. It's the collection of all vectors that can be formed from a given set. A basis, on the other hand, is like a team of robots chosen to explore the entire playground efficiently. In robotics, these concepts help us understand the space of possible motions and configurations.

Coordinate Systems: Robots often operate in different coordinate systems. These systems are like different languages, each with its own rules. Linear algebra helps robots smoothly translate between coordinate systems, ensuring they can navigate and interact in various environments.

Homogeneous Coordinates: Homogeneous coordinates are like a universal translator for robots. They extend the regular coordinates to handle translations and rotations seamlessly. This is crucial when robots need to move and orient themselves accurately in 3D space.

Transformation Matrices: Transformation matrices are like the master scripts of robot movements. They combine translations and rotations into a single matrix, providing a unified way to express complex transformations. Imagine a robot assembling parts—a transformation matrix guides its every move.

Linear Systems: Robots often encounter linear systems of equations, which are like puzzle pieces that need to fit together. Linear algebra helps robots solve these equations to determine unknowns, such as joint angles or forces. It's like a robotic detective uncovering the missing pieces of information.

Least Squares Solutions: In real-world scenarios, not all problems have exact solutions. Least squares solutions are like robots finding the best-fit solutions when faced with noisy or incomplete data. This is useful in applications like robotic perception, where sensors might provide imperfect information.

Optimization: Optimization is like a robot seeking the best path to achieve a goal. Linear algebra provides tools for optimizing robot movements and control strategies. For instance, a robot planning its trajectory through a cluttered environment uses optimization to find the most efficient path.

Robot Kinematics: Robot kinematics is the study of robot motion without considering forces. It's like understanding the geometry of a robot's movements. Linear algebra plays a central role in kinematics, enabling robots to calculate their joint angles, end-effector positions, and orientations.

Robot Dynamics: Robot dynamics, on the other hand, considers the forces and torques involved in robot motion. It's like analyzing the physics of a robot's movements. Linear algebra helps robots model and simulate dynamic behaviors, ensuring they can interact with their environment safely and effectively.

Localization and Mapping: In robotics, localization and mapping are essential for robots to navigate and understand their surroundings. Linear algebra aids in estimating a robot's position and creating maps of its environment using sensor data. Imagine a robot exploring an unknown area—it uses localization and mapping to know where it is and what's around it.

Control Systems: Control systems are like the orchestra conductors of robots' movements. Linear algebra provides tools for designing control algorithms that regulate a robot's actions. Whether it's maintaining balance, tracking a trajectory, or grasping an object, control systems ensure robots perform tasks accurately and efficiently.

Inverse Kinematics: For robots with multiple joints, inverse kinematics is like solving a puzzle in reverse. It determines the joint angles needed to achieve a desired end-effector position and orientation. This is crucial when a robot needs precise control over its movements, such as in robotic arms or humanoid robots.

Path Planning: Path planning is like plotting a roadmap for robots. It involves finding a sequence of movements that take a robot from its current position to a goal while avoiding obstacles. Linear algebra helps in the calculations necessary for smooth and obstacle-free navigation.

Machine Learning and Robotics: Linear algebra forms the foundation of machine learning algorithms, which are increasingly integrated into robotics. Robots use machine learning to adapt to changing environments, recognize objects, and make decisions based on data. It's like giving robots the ability to learn and improve from experience.

Simulation and Modeling: Before robots embark on physical missions, they often undergo simulations. These simulations use linear algebra to model robot behaviors and test different scenarios. It's like a virtual rehearsal for robots, allowing engineers to fine-tune their algorithms and strategies.

Error Analysis: In the real world, robots encounter errors in their measurements and movements. Linear algebra aids in error analysis, helping robots understand the accuracy and uncertainty of their actions. It's like robots becoming aware of their limitations and making adjustments to compensate.

Robotics Research and Innovation: Linear algebra is at the forefront of robotics research and innovation. It enables the development of advanced control algorithms, perception systems, and autonomous capabilities. With each new discovery and breakthrough in linear algebra, robots become more capable and versatile, pushing the boundaries of what they can achieve.

Ethical Considerations: In the ever-expanding field of robotics, ethical considerations are paramount. As robots become more integrated into our lives, linear algebra plays a role in ensuring that they are used responsibly and in ways that benefit society as a whole.

Linear algebra, is the mathematical backbone that empowers robots to understand their world, move with precision, and interact with humans and their environment. It's the invisible force that makes robotics not just a science fiction dream but a reality that continues to evolve and shape our future.

Chapter 3: Essential Mathematics for Robotics

Calculus and Differential Equations in Robotics
Imagine we're diving into a thrilling adventure where the language of mathematics becomes the bridge between the conceptual and the practical in robotics.

Calculus as the Language of Change: At the heart of calculus lies the concept of change, and change is a fundamental aspect of robotics. Calculus allows us to understand and describe how things change over time. Think of a robot's movement: its position, velocity, and acceleration. Calculus enables us to precisely analyze and control these dynamic aspects.

Derivatives: Derivatives are like the detective tools of calculus. They help us understand the rate of change of a quantity. When applied to robotics, derivatives allow us to determine how fast a robot is moving or how quickly its sensors are detecting changes in the environment. For instance, imagine a self-driving car using derivatives to calculate its speed and adjust its course to navigate a curve smoothly.

Integrals: Integrals are like the accumulators of calculus. They help us sum up quantities over time or space. In robotics, integrals play a pivotal role in tasks like mapping an environment or calculating the total distance traveled. Imagine a robot exploring an area and using integrals to create a map by integrating information from its sensors over time.

Velocity and Acceleration: For robots, velocity and acceleration are crucial. Velocity tells us how fast a robot is moving, while acceleration reveals how quickly its speed is changing. Calculus allows us to compute and control these quantities, ensuring that robots move smoothly and make precise adjustments when needed.

Kinematics: Kinematics in robotics deals with the geometry of motion. It's like understanding the dance steps of a robot. Calculus helps us describe the relationships between positions, velocities, and accelerations of robot parts, such as robotic arms or wheels. This knowledge is essential for controlling and planning robot movements.

Path Planning: Imagine a robot navigating a cluttered environment. Path planning involves finding a safe and efficient route from point A to point B while avoiding obstacles. Calculus aids in determining the optimal path by considering the robot's dynamics and the environment's constraints.

Differential Equations: Differential equations are the equations that describe how quantities change continuously. They're like the dynamic scripts that govern a robot's behavior. In robotics, differential equations model various aspects, from the motion of robot joints to the behavior of control systems. These equations allow us to predict and control robot responses accurately.

Forward Kinematics: Forward kinematics is the process of determining the position and orientation of a robot's end-effector (such as a gripper) based on the joint angles. Differential equations are used to model the transformations between different parts of the robot, helping us understand how changes in joint angles affect the end-effector's position.

Inverse Kinematics: Inverse kinematics, on the other hand, involves finding the joint angles required to achieve a desired end-effector position and orientation. Differential equations are instrumental in solving this complex puzzle, allowing robots to perform precise and coordinated movements, like picking up an object with a robotic arm.

Control Systems: Control systems in robotics are responsible for regulating a robot's behavior. They're like the conductors orchestrating a robot's movements. Differential equations play a pivotal role in control systems by modeling the dynamics of the robot and the environment. This modeling allows controllers to make real-time decisions to achieve desired behaviors.

Robot Dynamics: Robot dynamics delve into the forces and torques involved in robot motion. It's like understanding the physical forces at play when a robot interacts with its surroundings. Differential equations describe how these forces affect a robot's movement, ensuring it can operate safely and effectively.

Trajectory Planning: Robots often follow specific trajectories, whether it's a drone's flight path or a robotic arm's motion. Differential equations help in planning and executing these trajectories. For instance, think of a drone following a curved path—it relies on differential equations to adjust its thrust and orientation continuously.

Optimization: Optimization is like finding the best strategy for a robot to achieve its goals. Differential equations come into play when optimizing robot movements, allowing us to minimize energy consumption, maximize speed, or achieve other objectives efficiently.

Machine Learning Integration: In the era of AI and machine learning, differential equations are integrated into learning algorithms. Robots use machine learning techniques to adapt to their environments. Differential equations help in modeling and simulating dynamic systems, allowing robots to learn and make decisions based on real-time data.

Sensor Fusion: Robots often use multiple sensors to perceive their surroundings. Sensor fusion, which involves combining data from various sensors, relies on differential equations to create a coherent and accurate representation of the environment. This enables robots to make informed decisions, such as autonomous navigation in unpredictable environments.

Localization and Mapping (SLAM): In the realm of simultaneous localization and mapping (SLAM), robots use differential equations to estimate their position and create maps of their environment. It's like a robot's self-awareness and mapping abilities, crucial for applications like autonomous exploration and search and rescue.

Real-Time Decision-Making: Robots operate in real-time environments where split-second decisions are crucial. Differential equations enable robots to predict future states, assess risks, and make decisions in real-time. Consider a robot navigating a busy street—it uses differential equations to calculate safe trajectories and avoid collisions.

Simulation and Testing: Before robots are deployed in the real world, they undergo extensive simulations. Differential equations play a key role in modeling the robot's behavior and interactions, ensuring that it performs as expected in various scenarios.

Error Analysis: Robots encounter errors in their measurements and actions. Differential equations help in error analysis, allowing robots to assess the accuracy and uncertainty of their observations and movements. This is vital for safe and reliable robot operations.

Advanced Control Strategies: As robotics advances, more sophisticated control strategies are developed. Differential equations provide the mathematical foundation for these strategies, allowing robots to exhibit complex behaviors, such as cooperative tasks among multiple robots or human-robot collaboration.

Ethical Considerations: In the rapidly evolving field of robotics, ethical considerations are paramount. The use of calculus and differential equations in robotics must align with ethical principles, ensuring that robots are designed and operated in ways that prioritize safety, fairness, and societal benefit.

In summary, calculus and differential equations are the mathematical engines driving robotics forward. They empower robots to perceive the world, make intelligent decisions, and move with precision. From the simplest tasks to the most complex missions, the language of mathematics continues to expand the horizons of what robots can achieve, promising a future where they become indispensable partners in various aspects of our lives.

In our journey through robotics, we've uncovered the importance of mathematics as the language that bridges the gap between imagination and reality. Probability and statistics, in particular, are the versatile tools that enable robots to navigate uncertainty and make informed decisions in dynamic environments.

Imagine a robot embarking on a mission in an ever-changing world. It encounters unforeseen challenges, unexpected variations, and countless unknowns. This is where probability and statistics step in as the guiding stars, allowing the robot to adapt and thrive.

Probability, the art of quantifying uncertainty, serves as the foundation. Think of it as the robot's way of assigning chances to different outcomes. For instance, when a robot senses an object, it uses probability to estimate the object's position and uncertainty.

In robotics, probability takes many forms, but one of the most fundamental is Bayesian probability. It's like the robot's inner compass, helping it update its beliefs as new information arrives. Bayesian probability allows the robot to fuse data from various sensors, refining its understanding of the environment with each observation.

Uncertainty is a constant companion in the world of robotics. Sensors can be noisy, and the environment can be unpredictable. This is where statistics comes into play. Statistics is like the robot's statistical detective, helping it discern meaningful patterns from noisy data.

Imagine a robot tasked with recognizing objects from images. It uses statistical techniques to learn from a dataset of images and extract features that distinguish one object from another. These statistical patterns become the basis for its object recognition capabilities.

Regression analysis, a statistical method, is like the robot's way of finding the best-fit line through a scatterplot of data. For example, if a robot needs to predict the future position of a moving object, regression analysis allows it to make educated guesses based on past observations.

In robotics, the Kalman filter is a powerful tool that combines probability and statistics to estimate the state of a dynamic system. It's like the robot's ability to predict where things will be in the future based on their past movements. For instance, a drone uses the Kalman filter to estimate its position and velocity, even when GPS signals are unreliable.

Probability distributions are like the robot's toolkit for modeling uncertainty. The Gaussian distribution, also known as the normal distribution, is a common choice. Imagine a robot trying to predict the next location of a moving target. It uses a Gaussian distribution to represent the uncertainty in the target's position.

Robot localization, the process of determining a robot's position in an environment, relies heavily on probability and statistics. It's like the robot's GPS system, allowing it to know where it is relative to its surroundings. Monte Carlo localization is one popular technique, where the robot uses random sampling to estimate its position.

When robots operate in uncertain environments, they need to plan their actions carefully. Path planning under uncertainty is like the robot's strategic chess game, where it considers different possible outcomes and selects the path that minimizes risks. Here, probability and statistics help the robot make decisions that balance exploration and exploitation.

Imagine a robot exploring the ocean depths. It encounters unknown underwater terrain with hidden obstacles. To navigate safely, the robot uses probabilistic mapping. This technique allows it to build a map of the environment while considering the uncertainty in its sensor measurements.

Occupancy grids are a common representation in robotics, akin to the robot's blueprint of its surroundings. They divide the environment into a grid, with each cell representing the probability of occupancy. Robots use occupancy grids for tasks like obstacle avoidance and mapping in unknown environments.

Simultaneous Localization and Mapping (SLAM) is like the robot's cartographer and geographer rolled into one. It's the art of a robot creating a map of its environment while simultaneously determining its own position within that map. SLAM uses probabilistic methods to handle the inherent uncertainty in both mapping and localization.

Machine learning and artificial intelligence (AI) have become integral to robotics, and probability and statistics are at the core of these technologies. Consider a robot learning to recognize hand gestures. It uses probabilistic models to distinguish between different gestures, adapting and improving its accuracy over time.

Reinforcement learning, a branch of AI, is like the robot's teacher. It learns by trial and error, using probability to estimate the likelihood of different actions leading to rewards. Robots employ reinforcement learning to master tasks like playing games or controlling complex systems.

In the realm of autonomous driving, robots must make decisions in real-time while considering the uncertainty of the road environment. They use probabilistic algorithms to predict the movements of other vehicles, ensuring safe and efficient navigation through traffic.

Robotic perception relies heavily on probability and statistics. When a robot processes sensor data to understand its surroundings, it uses techniques like Bayesian filtering. This allows it to estimate the true state of the world, even when sensors provide uncertain or conflicting information.

Imagine a robot exploring a disaster-stricken area. It uses probabilistic reasoning to identify areas with a higher likelihood of survivors, guiding rescue efforts effectively. Probabilistic models help prioritize search and rescue operations in complex and uncertain scenarios.

Ethical considerations in robotics are paramount, and probability and statistics play a role here too. When robots make decisions that affect human lives, it's essential to ensure that these decisions are fair and unbiased. Probabilistic fairness algorithms aim to address issues of discrimination and bias in decision-making by robots.

As robots become more integrated into our lives, trust becomes a crucial factor. Probability and statistics provide the tools for quantifying and conveying uncertainty to users. Robots can communicate their confidence levels in decisions, allowing humans to make informed judgments about the robot's actions.

In summary, probability and statistics are the guiding lights in the world of robotics. They empower robots to embrace uncertainty, adapt to dynamic environments, and make decisions that are not only efficient but also ethical and fair. In this ever-evolving field, the marriage of mathematics and robotics continues to redefine what robots can accomplish, promising a future where they are not just machines but trusted partners in various aspects of our lives.

Chapter 4: Programming Robots: Getting Started

Imagine stepping into a workshop filled with robots of various shapes and sizes, each waiting to be infused with intelligence and purpose. Robot programming languages are the magic wands that allow us to communicate our intentions to these mechanical wonders.

At the heart of this fascinating realm lies the need to bridge the gap between human understanding and robot actions. Robot programming languages serve as the translators, converting our ideas and instructions into a language that robots comprehend.

While programming languages for computers have been around for decades, robot programming languages are specialized tools designed to cater to the unique needs of robots. They consider the physicality and dynamics of robots, their sensory capabilities, and the real-world environments in which they operate.

Robot programming languages are like the secret codes that unlock a robot's potential. They provide a structured way to specify what a robot should do, how it should do it, and under what conditions. This structured approach is essential for developing complex robotic systems.

One of the primary challenges in robot programming is the need for robots to interact with the physical world. This interaction requires precise control over a robot's movements and actions. Robot programming languages provide the means to specify motion trajectories, coordinate movements of robot limbs, and manipulate objects with precision.

In the world of robotics, there is no one-size-fits-all programming language. Instead, a variety of languages exist, each tailored to specific tasks, platforms, and robot types. These languages can be broadly categorized into two main types: low-level and high-level.

Low-level robot programming languages are like the nuts and bolts of a robot's operation. They provide fine-grained control over hardware components, such as motors, sensors, and actuators. These languages are often used for tasks that demand precise control and real-time responsiveness.

Assembly languages, for instance, allow programmers to interact directly with a robot's hardware. They are like the blueprint of a robot's inner workings, enabling programmers to manipulate memory, registers, and control signals. Assembly languages are essential for tasks where timing and hardware control are critical.

C and C++ are also commonly used for low-level robot programming. They provide a balance between control and portability, making them suitable for embedded systems and real-time applications. Programmers can use these languages to optimize robot performance while maintaining a level of abstraction.

High-level robot programming languages, on the other hand, are like the orchestras that play beautiful melodies with ease. They offer a more abstract way of describing robot behaviors, allowing programmers to focus on tasks rather than low-level hardware details.

Python is a popular high-level language in the world of robotics. Its simplicity and readability make it an excellent choice for prototyping and experimenting with robots. Python libraries and frameworks cater to various robotic applications, from computer vision to machine learning.

ROS (Robot Operating System) is a powerful framework that provides a standardized way to develop and control robots. It offers a suite of tools, libraries, and communication protocols that simplify the development of robotic systems. ROS uses a combination of C++ and Python, making it accessible to a wide range of developers.

MATLAB is another versatile tool often used in robotics research and development. Its rich ecosystem of toolboxes and libraries allows engineers and researchers to model, simulate, and control robotic systems efficiently. MATLAB is particularly popular for tasks involving mathematical modeling and control design.

Scratch, a visual programming language, introduces the world of robotics to young and aspiring programmers. It uses a block-based interface, making it accessible to children and beginners. With Scratch, users can program robots by snapping together code blocks, fostering creativity and learning through experimentation.

Blockly, a web-based visual programming language, shares similarities with Scratch but is tailored for robotics and IoT (Internet of Things) applications. It allows users to create code by stacking blocks that represent various functions and commands. Blockly is a user-friendly entry point into robotics programming.

In the context of industrial robots, robot programming languages like RAPID (Robot Application Programming Interface for Developers) and KRL (KUKA Robot Language) are used. These languages provide manufacturers and engineers with the means to program and control robots in manufacturing and assembly lines.

The versatility of robot programming languages extends beyond the programming itself. They enable robots to interact with a wide range of sensors, from cameras and LIDAR (Light Detection and Ranging) to touch sensors and proximity detectors. These languages allow robots to make sense of their surroundings, detect objects, and respond to changes in their environment.

Robot programming languages are also instrumental in human-robot interaction (HRI). They facilitate the development of interfaces that allow users to control robots through natural language commands, gestures, or even brain-computer interfaces. This capability opens up exciting possibilities for collaborative and assistive robotics.

The concept of robot programming goes beyond traditional coding; it encompasses various paradigms and methodologies. Behavior-based programming, for instance, focuses on defining behaviors or skills that a robot should exhibit. These behaviors can be combined and orchestrated to achieve complex tasks, much like composing a symphony.

Planning and control algorithms play a crucial role in robot programming. Path planning algorithms enable robots to navigate through cluttered environments, avoid obstacles, and reach their destinations efficiently. These algorithms are essential for mobile robots, autonomous vehicles, and drones.

Inverse kinematics, a mathematical technique, is often employed in robot programming to determine the joint angles needed to achieve a specific end-effector position and orientation. This is particularly important for tasks that require precise manipulation, such as pick-and-place operations in manufacturing.

Machine learning and AI techniques have also found their way into robot programming. Reinforcement learning, for instance, allows robots to learn and adapt their behaviors through interaction with the environment. Neural networks can be trained to recognize objects, interpret natural language, and make decisions.

Simulation tools are indispensable in robot programming. They provide a safe and efficient environment for testing and validating robot behaviors before deploying them in the real world. These simulations help reduce development time and minimize the risk of costly errors.

The future of robot programming languages is exciting and full of possibilities. As robots become more integrated into our lives, there is a growing demand for intuitive and accessible programming interfaces. Natural language interfaces, augmented reality (AR) programming, and cloud-based development platforms are some of the emerging trends in this field.

In summary, robot programming languages are the versatile conduits that allow us to breathe life into robots, enabling them to navigate, interact, and serve us in diverse ways. Whether you're orchestrating the precise movements of an industrial robot arm or teaching a robot to recognize faces and respond to voice commands, these languages are the keys to unlocking the potential of robotics. As technology continues to advance, the boundaries of what robots can achieve are expanding, promising a future where robots become indispensable partners in various aspects of our lives.

Imagine the excitement of embarking on a journey to create your first robot program. It's a moment filled with possibilities and the promise of bringing a mechanical friend to life. In this chapter, we'll explore the steps and concepts involved in crafting your very own robot program.

At the heart of robot programming is the desire to instruct a machine to perform specific tasks autonomously. It's like giving a robot a set of skills and abilities to navigate its environment, make decisions, and interact with the world around it.

Before diving into coding, it's essential to have a clear understanding of your robot's capabilities and objectives. Like planning a road trip, defining your destination and route is crucial. What do you want your robot to achieve? Is it a simple task like moving from point A to point B, or does it involve more complex interactions?

Selecting the right programming language for your robot is akin to choosing the language for communication. Depending on your robot's platform and your familiarity with programming languages, you might opt for a high-level language like Python or a specialized language like ROS for robotics.

Understanding your robot's hardware is fundamental. Just as you'd need to know the type of engine in your car, you should know your robot's sensors, actuators, and control systems.

These components are the tools your robot uses to perceive its environment and execute actions.

Robot programming often involves a combination of motion control and sensor integration. Think of it as teaching your robot to sense the world and react accordingly. For example, if your robot has a camera, you can program it to recognize objects or faces.

Designing your program's architecture is like planning the structure of a building. Will it be a simple, linear sequence of actions, or will it involve more complex decision-making and branching? Object-oriented programming concepts can help organize your code effectively.

Debugging is a crucial skill in robot programming. Just as you'd troubleshoot your car's engine, you'll need to identify and fix issues in your robot's code. Debugging tools and techniques, like print statements and debugging environments, will be your trusty companions.

Robots often operate in real-world environments where unexpected situations can arise. It's essential to incorporate error handling and exception handling into your program. This is like preparing for unexpected detours during your road trip.

Testing your robot program is like taking a test drive. You'll want to ensure that your code behaves as expected in different scenarios. Simulation tools can be invaluable for testing and refining your program before deploying it on the actual robot.

Integration with sensors is a critical aspect of robot programming. Imagine your robot's sensors as its eyes and ears. They provide data about the environment, such as distance measurements, camera images, or temperature readings. Your program needs to process this data effectively.

Actuators, on the other hand, are the robot's muscles. They enable your robot to perform physical actions, such as moving wheels, opening grippers, or controlling motors. Programming these actuators allows your robot to translate decisions into movements.

Kinematics, a branch of mathematics, plays a role in robot programming, especially for robotic arms or manipulators. It's like teaching your robot the rules of its own body, enabling it to calculate joint angles and end-effector positions.

Path planning is akin to charting a course for your robot. It involves determining the best route or trajectory for your robot to follow while avoiding obstacles. This is crucial for navigation tasks, whether your robot is exploring unknown terrain or moving objects in a factory.

Control algorithms are the brains behind your robot's actions. They determine how the robot should behave based on sensor input and its current state. Proportional-Integral-Derivative (PID) controllers, for instance, are commonly used for tasks like maintaining a robot's position.

Creating a graphical user interface (GUI) for your robot program can enhance user interaction. It's like designing a dashboard for your car. A well-designed GUI allows users to monitor the robot's status, provide input, and receive feedback.

Machine learning and AI techniques can elevate your robot's capabilities. Imagine teaching your robot to recognize handwritten characters or adapt to changing environments. Integrating machine learning models into your program allows your robot to learn from data and improve over time.

Networking and communication protocols enable your robot to connect with other devices and systems. Just as your car can communicate with your smartphone, your robot can exchange data with other robots, computers, or the cloud. This connectivity expands your robot's capabilities.

Security considerations are vital in robot programming, especially when robots are used in critical applications. Just as you'd protect your car from theft, you need to safeguard your robot from unauthorized access and potential cybersecurity threats.

Documentation is like the owner's manual for your robot program. It's essential to document your code, algorithms, and system architecture. This documentation not only helps you understand and maintain your program but also enables others to work with it.

Collaboration and community support can be invaluable in your robot programming journey. Just as you might seek advice from fellow travelers on a road trip, engaging with the robotics community, participating in forums, and attending conferences can provide guidance and inspiration.

As you embark on creating your first robot program, remember that it's a journey of exploration and learning. Like any adventure, there may be challenges and unexpected twists along the way. Embrace them as opportunities to grow and enhance your robot programming skills.

The satisfaction of seeing your robot come to life and perform tasks you've programmed is akin to reaching your destination after a long road trip. It's a testament to your creativity and technical prowess, and it opens doors to a world of possibilities in robotics.

In summary, creating your first robot program is a thrilling endeavor that combines creativity, technical skills, and problem-solving. Just as you'd carefully plan a trip, take the time to understand your robot's capabilities, choose the right tools, and design a program that brings your robot's abilities to life. Whether your robot is a small educational platform or a sophisticated industrial machine, your programming skills are the key to unlocking its potential and enabling it to contribute to the world of robotics.

Chapter 5: Sensors and Perception

In the intricate world of robotics, sensors are the eyes and ears of machines, enabling them to perceive and interact with the environment. These sensors come in various types, each with its unique abilities and applications, much like the diverse senses of living creatures.

Imagine a robot standing in a room, ready to explore and interact with its surroundings. To do so effectively, it relies on sensors to gather information about the world around it. These sensors are its windows to the outside, helping it navigate, make decisions, and perform tasks.

One of the most common types of sensors in robotics is the proximity sensor. These sensors are like the robot's personal space detectors, capable of detecting the presence or absence of objects in their vicinity. Infrared proximity sensors, for instance, emit infrared light and measure the reflection to determine the distance to nearby objects.

Ultrasonic sensors, another type of proximity sensor, use sound waves to detect obstacles. They work on the principle of sending out sound pulses and measuring the time it takes for the sound to bounce back. Ultrasonic sensors are like the robot's echolocation system, enabling it to navigate by detecting objects and obstacles.

Vision sensors are akin to a robot's eyes, allowing it to perceive the world through images and videos. Cameras are the most common vision sensors, capturing visual information that can be processed for object recognition, navigation, and even facial expression analysis. Vision sensors are essential for tasks like autonomous navigation and object detection.

LIDAR (Light Detection and Ranging) sensors are the high-precision eyes of robotics. They emit laser beams and measure the time it takes for the light to bounce back from objects. LIDAR sensors provide detailed 3D maps of the environment, enabling robots to navigate complex terrain, avoid obstacles, and even perform simultaneous localization and mapping (SLAM).

Infrared sensors are like the robot's heat detectors, capable of sensing thermal radiation. They are often used for applications like temperature monitoring, object tracking, and detecting the presence of humans or animals based on their body heat. Infrared sensors are valuable in home automation and security systems.

Force and pressure sensors allow robots to perceive and respond to physical interactions with their environment. They are like a robot's sense of touch, providing feedback on the force or pressure applied to their end-effectors or surfaces. These sensors are crucial in applications such as robotic grippers, prosthetics, and manufacturing tasks that require delicate handling.

Tactile sensors are specialized sensors designed to mimic the human sense of touch. They are like the robot's sensitive skin, capable of detecting pressure, vibration, and even texture. Tactile sensors enable robots to manipulate objects with precision and interact with the world in a more nuanced way.

Accelerometers and gyroscopes are motion sensors that help robots understand their orientation and movement. Accelerometers measure linear acceleration, while gyroscopes measure angular velocity. These sensors are like a robot's inner ear, allowing it to maintain balance and stability, essential for tasks like self-balancing robots and drones.

Magnetic sensors are the compasses of robotics, helping robots navigate by detecting magnetic fields. They are valuable in applications like autonomous vehicles and drones, where knowing the direction is crucial for navigation and orientation.

Environmental sensors, such as temperature sensors, humidity sensors, and gas sensors, provide information about the robot's surroundings. They are like the robot's weather station, allowing it to monitor environmental conditions and make decisions accordingly. Environmental sensors are essential in applications like environmental monitoring and indoor climate control.

Sound sensors, or microphones, enable robots to perceive auditory information from their environment. They are like a robot's ears, capable of capturing sounds and speech. Sound sensors are valuable in human-robot interaction, voice recognition, and surveillance applications.

Chemical sensors detect specific chemicals or gases in the environment. These sensors are like the robot's nose, allowing it to identify and respond to chemical substances. Chemical sensors have applications in areas such as industrial safety, environmental monitoring, and detecting gas leaks.

Biological sensors, although less common, are used in some specialized robotics applications. These sensors can detect biological signals such as electrical impulses in muscles or brain activity. They are like the robot's biological interfaces, enabling direct interaction with biological systems in fields like medical robotics and brain-computer interfaces.

As robots become more integrated into our lives, sensor fusion has become a crucial concept. It involves combining data from multiple sensors to enhance a robot's perception and decision-making abilities. Sensor fusion is like the robot's ability to integrate information from its various senses to form a coherent understanding of the world.

Machine learning and artificial intelligence are often used in conjunction with sensors to interpret and process sensor data. Robots can be trained to recognize patterns in sensor data, allowing them to make informed decisions and adapt to changing environments.

Sensor calibration is an essential step in ensuring the accuracy and reliability of sensor measurements. Just as we need to calibrate a compass or a thermometer, robots require calibration to correct for any inaccuracies or biases in their sensors.

In the grand symphony of robotics, sensors are the orchestra, playing a vital role in enabling robots to perceive, interact, and adapt to their environment. These diverse sensors provide robots with a range of capabilities, from detecting objects and navigating terrain to responding to touch and sound. As technology advances, the fusion of sensor data with AI and machine learning promises to elevate the sensory experiences of robots, making them even more capable and adaptable companions in our ever-evolving world.

In the world of robotics, sensor data processing and fusion are the secret sauce that allows robots to make sense of the complex information gathered by their various sensors. Think of it as the brain behind the senses, enabling robots to perceive the world, make decisions, and act with intelligence.

Imagine a robot navigating a cluttered room, filled with objects, people, and obstacles. It relies on its sensors, much like our senses of sight, hearing, and touch, to gather information about its surroundings. However, raw sensor data can be noisy and overwhelming, much like a cacophony of sounds. This is where sensor data processing and fusion come into play.

Sensor data processing is like the robot's ability to filter out irrelevant noise from its sensory input. It involves techniques to clean, preprocess, and enhance raw sensor data. For example, image processing can remove image artifacts, while signal processing can filter out background noise from auditory data. The goal is to ensure that the data accurately represents the robot's environment.

Feature extraction is another aspect of sensor data processing. It's like the robot's ability to identify distinct patterns or features in the data. For instance, in image data, feature extraction can identify edges, shapes, and objects. In audio data, it can extract spectral features for speech recognition. These extracted features serve as the building blocks for higher-level analysis.

Data fusion is the art of combining information from multiple sensors to create a more comprehensive and accurate representation of the environment. Imagine the robot's sensors as a group of friends sharing their observations. Each friend has a unique perspective, but by combining their insights, a clearer picture emerges.

Sensor fusion can be categorized into two main types: sensor-level fusion and decision-level fusion. Sensor-level fusion involves merging raw sensor data, while decision-level fusion combines the outputs of individual sensors' processing algorithms. It's like assembling a jigsaw puzzle where each sensor contributes a piece to complete the picture.

Sensor-level fusion includes techniques like data alignment and registration. It's like ensuring that all friends speak the same language and have the same frame of reference. Data alignment ensures that sensor data from different sources are synchronized and aligned spatially and temporally.

Kalman filters are often used in sensor-level fusion to estimate a system's state based on noisy sensor measurements. Imagine the robot as a detective piecing together clues from various sources to track the movement of an object. Kalman filters help maintain a consistent and accurate estimate of the object's position and velocity.

Decision-level fusion involves making decisions based on the outputs of individual sensors. It's like having a group of friends vote on a decision, with each friend providing their opinion. The final decision is made by combining these opinions using techniques like voting algorithms or weighted averaging.

Multi-sensor fusion allows robots to compensate for the limitations of individual sensors. Just as we trust multiple senses to perceive the world, robots rely on a combination of sensors to make informed decisions. For example, a robot can use both visual and depth sensors to detect and avoid obstacles while navigating.

Localization and mapping are critical applications of sensor data processing and fusion. Imagine the robot as an explorer mapping an uncharted territory. It combines data from sensors like LIDAR, cameras, and IMUs (Inertial Measurement Units) to create a detailed map of its surroundings while simultaneously determining its own position within that map.

Simultaneous Localization and Mapping (SLAM) is a key technique in robotics that combines localization and mapping in real time. It's like the robot's ability to create a map while knowing its precise location within that map. SLAM algorithms use sensor data to estimate the robot's pose (position and orientation) and build a map of the environment simultaneously.

Object recognition is another application of sensor data processing and fusion. Imagine the robot as a detective trying to identify objects in a room. By combining data from visual sensors and depth sensors, the robot can recognize objects, estimate their size and position, and interact with them accordingly.

Gesture recognition is a human-robot interaction application that relies on sensor data processing. It's like the robot's ability to understand gestures and commands from a human user. By processing data from cameras and depth sensors, the robot can interpret hand movements and gestures, enabling intuitive control.

Human tracking and interaction involve using sensor data to detect and interact with humans in the robot's vicinity. It's like the robot's ability to recognize and respond to people. Sensors like cameras and depth sensors are used to track the movement and gestures of humans, enabling natural and safe interactions.

Environmental monitoring is a crucial application in robotics, especially in fields like agriculture and environmental science. It's like the robot's role as an environmental scientist, collecting data about temperature, humidity, air quality, and soil conditions using various sensors. This data helps make informed decisions for tasks like irrigation and pest control.

Medical robotics also benefits from sensor data processing and fusion. Imagine the robot as a surgical assistant, working alongside a surgeon. By integrating data from various sensors, including cameras, force sensors, and haptic feedback devices, the robot can provide precise and safe assistance during surgeries.

Sensor data processing and fusion play a pivotal role in autonomous vehicles. Think of the robot as a self-driving car navigating busy streets. By fusing data from LIDAR, radar, cameras, and GPS, the vehicle can perceive its surroundings, detect obstacles, and make decisions in real time.

Machine learning and deep learning techniques are often combined with sensor data processing to enhance a robot's capabilities. It's like the robot's ability to learn from its experiences and adapt to different situations. Neural networks can be trained on sensor data to recognize patterns, objects, and behaviors, enabling robots to perform complex tasks.

In summary, sensor data processing and fusion are the cognitive abilities that empower robots to perceive, interpret, and interact with their environment. These techniques enable robots to make sense of the complex and noisy sensory data they receive, much like our brains process the myriad of sensory inputs we encounter daily. As technology continues to advance, the fusion of sensor data with artificial intelligence promises to create even more intelligent and adaptive robots, opening new frontiers in robotics applications and human-robot collaboration.

Chapter 6: Robot Kinematics and Dynamics

Imagine a robot moving gracefully, reaching out its arm to pick up an object, much like a dancer performing an intricate routine. This fluidity of motion in robotics is made possible by understanding robot kinematics, a fundamental concept that defines how robots move their joints and end-effectors in space.

Robot kinematics is like the language of motion for robots, enabling them to perform precise and coordinated movements. To comprehend this concept, let's break it down into two key components: forward kinematics and inverse kinematics.

Forward kinematics is the first step in understanding how robots move. Think of it as the robot's ability to answer the question, "Where is my hand or end-effector located in space?" This aspect of kinematics involves mapping the joint angles of a robot to the position and orientation of its end-effector, such as a gripper or tool.

Imagine a robotic arm with multiple joints, each with a specific angle. The forward kinematics equations allow us to calculate the position (x, y, z) and orientation (roll, pitch, yaw) of the end-effector based on the joint angles. It's like solving a puzzle to determine the robot's hand position in the three-dimensional world.

These forward kinematics equations are derived using geometric and trigonometric principles. For a simple robot with a few joints, the calculations can be relatively straightforward. However, for complex robots with multiple degrees of freedom, the math becomes more intricate, involving matrices and transformation equations.

Inverse kinematics, on the other hand, is like the robot's ability to answer the question, "What joint angles should I use to reach a specific position and orientation?" This aspect of kinematics is essential for planning and controlling robot movements effectively.

Imagine you want the robot to pick up an object at a particular location and orientation. Inverse kinematics allows you to determine the joint angles needed to achieve this desired end-effector position and orientation. It's like figuring out how to position your arm and hand to grab an object on a cluttered shelf.

Inverse kinematics can be a challenging problem, especially for robots with many degrees of freedom or complex joint configurations. Solving these equations often involves iterative numerical methods, such as the Jacobian matrix and gradient descent algorithms. These techniques allow robots to find the optimal joint angles to reach their target positions.

Understanding robot kinematics is not limited to arms and manipulators. Mobile robots, such as self-driving cars and drones, also rely on kinematic principles. For these robots, kinematics involve determining how their wheels or thrusters should move to achieve specific translations and rotations in space.

For mobile robots, forward kinematics calculate how the robot's wheels or thrusters contribute to its overall movement. It's like understanding how the individual dancers in a troupe contribute to the group's choreography. By knowing how each wheel or thruster affects motion, we can plan and control the robot's path.

Inverse kinematics for mobile robots is like planning a dance routine. It involves determining the wheel or thruster speeds needed to achieve a desired movement, such as going from one point to another while avoiding obstacles. These calculations allow robots to navigate complex environments and execute tasks efficiently.

Kinematic models are essential tools in robot control and trajectory planning. They provide a mathematical representation of a robot's motion capabilities, allowing us to simulate and predict how the robot will behave in different situations. These models are like the choreographer's notes, guiding the robot's movements.

In robotics, kinematic chains are used to represent the interconnected joints and links of a robot. Each joint is like a dancer in a sequence, and the links are the connections between them. By understanding the kinematic chain, we can analyze and control the robot's movements effectively.

Parallel manipulators are a specialized type of robot with multiple limbs or arms. These robots are like acrobatic performers, using their parallel structure to achieve high precision and agility. Understanding the kinematics of parallel manipulators is essential for their control and coordination.

Redundant robots are those with more degrees of freedom than necessary for a task. They are like skilled improvisational dancers who can adapt to different styles and movements. Kinematic redundancy allows these robots to optimize their movements for efficiency or dexterity.

Kinematic singularities are critical points in a robot's motion where it loses some degrees of freedom. Imagine a dancer striking a pose that limits their subsequent movements. Kinematic singularities can affect a robot's ability to reach certain positions or orientations and require careful planning to avoid.

In robotic surgery, kinematics play a crucial role in controlling surgical instruments with precision. Surgeons use robotic systems to perform minimally invasive procedures with accuracy and stability. Understanding the kinematics of these systems ensures safe and effective surgeries.

Kinematic analysis is also essential in animation and computer graphics. It's like creating lifelike movements for animated characters or virtual simulations. By applying kinematic principles, animators can make characters and objects move realistically in digital environments.

In summary, understanding robot kinematics is like deciphering the graceful movements of a dancer on stage. It's the foundation that allows robots to perform tasks with precision and agility, whether it's picking up objects, navigating complex environments, or executing intricate maneuvers. As robotics continues to advance, kinematic principles will remain at the core of creating robots that move with grace and purpose, enriching our lives and industries.

Picture a robotic arm gracefully reaching out to perform a delicate task, much like a skilled performer on stage. This seamless motion is a result of understanding the dynamics of robotic arms, a fascinating field that explores how these mechanical wonders move and interact with their surroundings.

Robotic arm dynamics are like the intricate choreography of a dance, involving precise calculations and control to achieve fluid and coordinated movements. To delve into this topic, let's explore the key aspects of robotic arm dynamics.

First, consider the joints of a robotic arm. These joints are like the joints in our own limbs, allowing the arm to bend, twist, and rotate. Each joint has its unique range of motion and properties, similar to how different parts of our body move in distinct ways.

The motion of a robotic arm is governed by the principles of classical mechanics, just as our bodies obey the laws of physics. Newton's laws of motion and principles of energy conservation play a significant role in understanding how forces and torques affect the arm's motion.

When a robotic arm moves, it experiences various forces and torques, much like the forces acting on a dancer as they perform. These forces come from external factors such as gravity, friction, and contact with objects. Dynamics equations are used to calculate how these forces influence the arm's motion.

Inverse dynamics is a critical concept in robotic arm dynamics. It's like unraveling the choreography of a dance routine by working backward. Inverse dynamics allows us to determine the forces and torques needed at each joint to achieve a desired arm movement. This is essential for robotic control and trajectory planning.

Forward dynamics, on the other hand, is like choreographing a dance routine step by step. It involves simulating the motion of a robotic arm given a set of forces and torques applied to its joints. Forward dynamics helps predict how the arm will move in response to control inputs.

Robotic arms often consist of multiple links connected by joints, forming what is known as a kinematic chain. Each link is like a segment of a dancer's body, and the joints are the flexible connections between them. The dynamics of each link and joint contribute to the overall motion of the arm.

Payload capacity is a crucial consideration in robotic arm design. It's like knowing a dancer's ability to lift and balance a partner during a performance. The arm's payload capacity determines how much weight it can carry and manipulate without compromising its stability or performance.

The concept of inertia is essential in understanding robotic arm dynamics. Just as a dancer's body has mass and moments of inertia, each part of the arm has its mass and inertia properties. These properties influence how the arm accelerates and responds to external forces.

Dynamic modeling involves creating mathematical models that describe how a robotic arm's motion changes over time. These models are like choreography notes, capturing the arm's behavior and allowing for simulations and analysis. Dynamic models help engineers design and control robotic arms effectively.

Control algorithms play a significant role in robotic arm dynamics. It's like a conductor guiding an orchestra to perform a symphony. Control algorithms ensure that the arm's movements are precise, smooth, and coordinated. They take into account the dynamics of the arm to achieve desired tasks.

Feedback control is like a dancer adjusting their movements based on the music and partner's cues. In robotic arm dynamics, feedback control relies on sensors to provide real-time information about the arm's position, velocity, and forces. This information is used to adjust the arm's motion and maintain accuracy.

Robotic arms are used in a wide range of applications, from manufacturing and assembly to medical surgery and research. It's like a versatile performer who can adapt to various roles and styles of dance. Understanding the dynamics of robotic arms is crucial for optimizing their performance in these diverse fields.

In manufacturing and assembly, robotic arms are like skilled workers on the production line. They can perform repetitive tasks with precision and consistency, increasing efficiency and product quality. Robotic arm dynamics ensure that these tasks are executed accurately.

In medical surgery, robotic arms are like the steady hands of a surgeon, capable of performing intricate procedures with precision and stability. The dynamics of these arms are critical for ensuring safe and successful surgeries, especially in minimally invasive procedures.

Research applications often involve using robotic arms to manipulate objects, conduct experiments, or interact with the environment. It's like a scientist's tool, extending their reach and capabilities. Understanding the dynamics of robotic arms allows researchers to design experiments and conduct studies effectively.

Mobile robots, such as robotic rovers and drones, also have dynamic characteristics to consider. These robots move through space much like dancers navigating a stage. Their dynamics involve understanding how propulsion, balance, and control influence their movements.

In space exploration, robotic arms are like the versatile tools of astronauts, capable of performing tasks in the challenging environment of space. The dynamics of these arms are essential for tasks such as repairing spacecraft, handling equipment, and collecting samples.

Robotic arms in underwater environments face unique challenges, much like a dancer adapting to perform in water. Understanding their dynamics is crucial for tasks such as underwater exploration, maintenance of underwater structures, and deep-sea research.

In collaborative robotics, where robots work alongside humans, understanding arm dynamics is vital for ensuring safety and coordination. These robots are like dance partners, requiring precise control and responsiveness to human actions.

As robotic technology continues to advance, the field of robotic arm dynamics plays a pivotal role in creating more capable and versatile robots. Engineers and researchers work together to unravel the intricacies of arm dynamics, much like choreographers and dancers collaborating to create breathtaking performances.

In summary, robotic arm dynamics are at the heart of the mesmerizing movements of robots in various fields. Just as dancers master their craft to perform with grace and precision, understanding the dynamics of robotic arms is essential for unlocking their full potential in manufacturing, healthcare, research, and beyond. As technology evolves, so too will our ability to choreograph the elegant motions of these mechanical performers, enriching our lives and advancing the world of robotics.

Chapter 7: Robot Control Systems

Imagine a conductor leading an orchestra, guiding each musician to play their part in harmony. Robot control is much like that conductor, orchestrating the movements and actions of a robot to perform tasks with precision and efficiency.

At its core, robot control is about making decisions and sending commands to a robot's actuators, much like a conductor directs musicians to produce beautiful music. These commands dictate how the robot's joints and motors should move to achieve a desired action, whether it's picking up an object, painting a car, or navigating a room.

There are various levels of robot control, each with its complexity and purpose. At the lowest level is hardware control, which involves managing the robot's physical components, such as motors and sensors. It's like ensuring that each instrument in the orchestra is properly tuned and functioning.

Motor control, a subset of hardware control, focuses on managing the movement and behavior of the robot's motors and actuators. It's like instructing a musician to play a specific note or sequence. Motor control ensures that the robot's joints move accurately and smoothly.

Sensor feedback is essential for effective robot control, much like a conductor relies on the musicians' performance to adjust the tempo and dynamics of a piece. Sensors provide real-time information about the robot's position, orientation, and environmental conditions. This feedback helps the control system make informed decisions.

Kinematic control is like choreographing a dance routine for the robot. It involves planning and controlling the robot's movements in a coordinated manner. For example, if a robot is tasked with painting a car, kinematic control determines how it should move its arm and end-effector to cover the entire surface evenly. Dynamic control goes a step further by considering the forces and torques acting on the robot's joints and links. It's like the conductor adjusting the orchestra's performance to match the acoustics of the concert hall. Dynamic control ensures that the robot's movements are not only accurate but also safe and efficient.

End-effector control focuses on the precise manipulation of objects or tools held by the robot's end-effector, such as a gripper or welding torch. It's like a musician playing a solo, requiring fine-tuned control and coordination. End-effector control is crucial for tasks like assembly, welding, or pick-and-place operations. Task-level control is like composing a symphony for the robot. It involves defining high-level objectives and goals for the robot to achieve. Instead of specifying individual joint angles or motor commands, task-level control allows the operator to describe what they want the robot to do, such as "pick up the red ball" or "paint the car door." Path planning is essential for mobile robots navigating through environments, much like a GPS guiding a driver on a road trip. Path planning algorithms calculate the optimal route for the robot to reach its destination while avoiding obstacles. It ensures that the robot moves safely and efficiently in its environment. Trajectory planning is like planning the precise movements of a ballet dancer. For tasks that require continuous and smooth motion, trajectory planning defines a path for the robot to follow over time. It takes into account factors like acceleration, velocity, and jerk to ensure that the robot's movements are graceful and controlled.

Feedback control is a fundamental concept in robot control, much like a conductor adjusting the tempo based on the orchestra's performance. It involves continuously monitoring the robot's state and making real-time adjustments to achieve desired behavior. Feedback control ensures that the robot can adapt to changing conditions and uncertainties.

Proportional-Integral-Derivative (PID) control is a widely used technique in robot control, similar to adjusting the volume on a sound system to maintain a constant level. PID control uses three components – proportional, integral, and derivative terms – to regulate the robot's behavior. It's like finding the right balance between accuracy and stability. Inverse kinematics is essential for controlling robots with multiple degrees of freedom, like a conductor orchestrating a symphony with many instruments. It's like solving a puzzle to determine the joint angles required to achieve a specific end-effector position and orientation. Inverse kinematics allows for precise control of complex robotic arms. Model-based control leverages mathematical models of the robot's dynamics to plan and execute movements, much like a composer using sheet music to guide musicians. These models capture the relationship between control inputs and robot behavior, enabling accurate and predictable motion. Learning-based control, similar to a musician practicing to improve their skills, involves training the robot to adapt and improve its performance over time. Machine learning techniques, such as reinforcement learning or neural networks, allow robots to learn from experience and optimize their control strategies.

In collaborative robotics, where humans and robots work together, safety is a top priority. It's like ensuring that the orchestra's performance doesn't harm the conductor or other musicians. Collaborative robot control includes safety features like collision detection and force limiting to protect humans working alongside the robot. Robot programming languages are like the sheet music that guides musicians in an orchestra. These languages provide a means of communicating commands to the robot's control system. Common robot programming languages include Blockly, ROS (Robot Operating System), and G-code, each tailored to specific tasks and applications. Teaching robots through demonstration is akin to a conductor showing musicians how to play a piece. With this approach, operators physically guide the robot through a task, and the robot learns by mimicking the demonstrated movements. It's an intuitive way to train robots for various tasks. Teleoperation allows humans to control robots remotely, much like a conductor leading a virtual orchestra. Operators use joysticks, haptic devices, or other interfaces to send commands to the robot from a distance. Teleoperation is valuable for tasks in hazardous or challenging environments.

Human-robot interaction is a crucial aspect of robot control, similar to musicians collaborating in an ensemble. It involves designing interfaces and control systems that enable natural and intuitive communication between humans and robots. This includes gestures, voice commands, and touch-based interfaces. In summary, robot control is the art of orchestrating the movements and actions of robots to perform tasks effectively and efficiently. Just as a conductor brings out the best in an orchestra, robot control ensures that robots execute their tasks with precision and grace. Whether it's manufacturing, healthcare, research, or any other field, the principles of robot control continue to evolve, shaping a future where robots and humans work together harmoniously.

Imagine driving a car, and as you steer, you make constant adjustments to keep it on the desired course. This intuitive process of maintaining control is akin to the concept of feedback control systems in robotics and engineering.

Feedback control is the art of continuously monitoring and adjusting a system's behavior based on measured information, much like a skilled driver who uses their senses to navigate smoothly. In the world of robotics, this technique plays a pivotal role in ensuring robots perform tasks accurately and with precision.

At its core, feedback control involves three key components: a system, a sensor, and a controller. The system is like the vehicle you're steering, and the controller is your brain, making decisions. The sensor, on the other hand, is like your eyes, providing information about the system's state.

Think of a thermostat in your home as a simple example of a feedback control system. The thermostat continuously measures the temperature in the room (sensor), compares it to the desired temperature (reference), and adjusts the heating or cooling system (controller) to maintain the desired comfort level (system response). It's an ongoing cycle of sensing, comparing, and adjusting.

In robotics, sensors are the eyes and ears of the system, collecting data about the robot's position, orientation, velocity, and environmental conditions. These sensors can include cameras, encoders, accelerometers, gyroscopes, and many others, depending on the robot's application.

Control algorithms, like the conductor of an orchestra, orchestrate the actions of the robot. These algorithms use sensor data to make real-time decisions about how the robot's actuators (motors, joints, or thrusters) should move. The goal is to achieve desired behaviors and tasks, whether it's picking up an object, following a path, or maintaining stability.

Proportional-Integral-Derivative (PID) control, a widely used technique, is like tuning an instrument to create harmonious music. PID controllers adjust the robot's actions by considering the current error (the difference between the desired state and the actual state), the accumulation of past errors, and the prediction of future errors. This fine-tuning ensures that the robot's movements are smooth and accurate.

Closed-loop control is a fundamental concept in feedback control, much like a musician adjusting their playing based on the sound they hear. In closed-loop control, the system continuously receives feedback from the sensors, allowing it to make real-time adjustments to achieve the desired outcome. It's a dynamic and responsive approach to control.

Feedforward control, on the other hand, is like a conductor leading an orchestra based solely on the sheet music. In feedforward control, the controller makes decisions based on a predicted model of the system's behavior, without relying on feedback from sensors. It's an anticipatory approach that aims to precompensate for known disturbances or uncertainties.

Robust control is essential in situations where external factors or uncertainties can affect the system's performance, similar to a musician adapting to unexpected changes in the environment. Robust control algorithms are designed to ensure that the system remains stable and performs adequately under a wide range of conditions.

Adaptive control, like a musician learning to adapt to different styles of music, involves continuously adjusting the control strategy based on changing circumstances. Adaptive controllers use feedback to learn and adapt to variations in the system's dynamics, making them versatile and resilient in dynamic environments.

In real-world robotic applications, disturbances and uncertainties are inevitable, like unexpected variations in music tempo. To address these challenges, observers and estimators are used to estimate the system's state or the effects of disturbances. These estimations help the controller make informed decisions and maintain control.

Model-based control is like a conductor following a carefully composed musical score. It involves creating mathematical models that describe the robot's dynamics and behavior. These models serve as a reference for the controller, enabling it to predict how the system will respond to different inputs and disturbances.

State-space control, similar to a conductor managing various sections of an orchestra, allows for a more comprehensive representation of the system's behavior. In state-space control, the system's state, represented as a set of variables, is used to describe its behavior. This approach is particularly useful for complex robotic systems with multiple states and inputs.

Optimal control, like a composer striving for the perfect musical composition, seeks to find the best control strategy to achieve a specific objective. It involves formulating an optimization problem to minimize or maximize a performance criterion, such as minimizing energy consumption or maximizing task efficiency. Optimal control techniques provide a systematic way to optimize robot behavior.

H-infinity control is like a conductor aiming for the best possible performance, even in the presence of disturbances and uncertainties. H-infinity control focuses on designing controllers that minimize the worst-case effect of disturbances on the system's performance. It's a robust approach that ensures stable control under adverse conditions.

In robotics, the concept of trajectory tracking is like choreographing a dance routine for the robot. Trajectory tracking involves following a predefined path or trajectory with precision. Control algorithms ensure that the robot's movements closely match the desired trajectory, whether it's for tasks like painting, welding, or assembly.

Nonlinear control is essential for robots with complex dynamics, much like conducting a symphony with intricate harmonies. Nonlinear control techniques are designed to handle systems with nonlinear behaviors, which can be challenging to control using linear methods. These techniques enable robots to perform tasks that require high agility and precision.

In summary, feedback control systems are the maestros of the robotic world, orchestrating the movements and actions of robots with finesse and precision. These systems continuously adapt and adjust, much like a conductor leading a symphony through a dynamic musical journey. In robotics, feedback control ensures that robots perform their tasks with accuracy, resilience, and versatility, contributing to a future where robots and humans collaborate seamlessly in various fields.

Chapter 8: Machine Learning in Robotics

Imagine a robot capable of learning from its experiences, adapting to new situations, and making decisions like a skilled chess player strategizing moves. This remarkable ability is the essence of machine learning for robots, a field that empowers machines to acquire knowledge and improve their performance over time.

Machine learning is like teaching a robot to be a student of its environment, allowing it to observe, analyze, and make decisions based on data. In this introduction, we'll explore the foundations of machine learning and how it transforms robots into intelligent and adaptive entities.

At its core, machine learning involves training a computer or robot to recognize patterns and make predictions, much like a coach helping an athlete refine their technique. It's about imparting the ability to learn from data, whether it's visual information from cameras, sensor readings, or other sources.

Supervised learning is one of the fundamental paradigms in machine learning, like teaching a robot to identify objects by showing it labeled examples. In supervised learning, a model is trained on a dataset that includes input data and corresponding output labels. The model learns to map inputs to outputs, allowing it to make predictions on new, unseen data.

Unsupervised learning, on the other hand, is like discovering hidden patterns in a collection of artworks. In unsupervised learning, the model explores data without explicit labels or guidance. It seeks to identify structures or groupings within the data, such as clustering similar objects together.

Reinforcement learning is akin to training a dog with rewards and punishments. In reinforcement learning, a robot learns by interacting with an environment and receiving feedback in the form of rewards or penalties. It strives to maximize cumulative rewards, leading to the discovery of optimal strategies and behaviors.

Machine learning models, like the characters in a story, come in various forms. Neural networks, inspired by the human brain, consist of interconnected nodes that process and transform data. Convolutional neural networks (CNNs) are specialized for tasks like image recognition, while recurrent neural networks (RNNs) excel in sequential data analysis.

Decision trees, like a flowchart guiding a decision-making process, represent a series of choices and outcomes. These models are interpretable and suitable for tasks involving classification or regression.

Ensemble methods, like assembling a team of experts, combine multiple models to improve performance and generalization. Random forests and gradient boosting are examples of ensemble techniques that enhance the accuracy and robustness of machine learning models.

The process of training a machine learning model is like nurturing a young plant. It begins with data collection, where relevant information is gathered and prepared for analysis. This data is the nourishment that fuels the model's growth.

Data preprocessing, much like cleaning and refining raw materials, involves tasks like data cleaning, feature engineering, and normalization. These steps ensure that the data is in a suitable format for training.

Feature selection is like choosing the essential ingredients for a recipe. It involves selecting the most relevant features or variables that contribute to the model's performance while discarding unnecessary or redundant ones.

The training phase is where the magic happens. It's like a teacher guiding a student through lessons and exercises. During training, the model learns to recognize patterns and make predictions by adjusting its internal parameters.

Loss functions, similar to grading assignments, quantify how well the model is performing. The goal is to minimize the loss, indicating that the model's predictions are as close as possible to the actual outcomes.

Optimization algorithms, like search strategies, fine-tune the model's parameters to minimize the loss function. These algorithms ensure that the model converges to an optimal state during training.

Validation and testing, much like quizzes and exams, evaluate the model's performance on new and unseen data. These steps assess how well the model generalizes its learning to make accurate predictions outside the training set.

Overfitting is a common challenge in machine learning, like a student memorizing answers without understanding the material. It occurs when a model learns to perform exceptionally well on the training data but fails to generalize to new data. Techniques like regularization and cross-validation help prevent overfitting.

Hyperparameter tuning, similar to adjusting settings on a musical instrument, involves optimizing the model's hyperparameters to achieve the best performance. Grid search and random search are methods used to find the optimal hyperparameter values.

Transfer learning is like leveraging knowledge from one subject to excel in another. In transfer learning, a pre-trained model is adapted for a new task, saving time and resources. It's particularly useful when limited data is available for the target task.

Once a machine learning model is trained and validated, it's ready to be deployed, like a graduate stepping into the real world. Deployment involves integrating the model into a robotic system, allowing it to make real-time decisions and perform tasks autonomously.

Real-time inference is the process of applying the trained model to make predictions on incoming data, much like a musician performing in front of an audience. In robotics, real-

74

time inference enables the robot to react and adapt to its environment as it processes sensor data.

Machine learning for robots extends beyond pattern recognition and prediction. It enables robots to navigate through complex environments, make decisions in uncertain situations, and interact with humans naturally.

Robotics applications benefit from machine learning in various ways. Autonomous vehicles, like self-driving cars, rely on machine learning for perception, navigation, and decision-making. Machine learning helps these vehicles recognize road signs, detect obstacles, and plan safe routes.

In healthcare, robots assist in surgeries and patient care. Machine learning enables robots to analyze medical images, such as X-rays and MRIs, to diagnose diseases and assist in treatment planning. Surgical robots use machine learning to enhance precision and reduce invasiveness.

In manufacturing, robots equipped with machine learning excel in tasks like quality control and defect detection. These robots can inspect products for defects, classify them, and make decisions on whether they meet quality standards.

Robotic assistants, like smart home devices, leverage machine learning to understand and respond to human commands. Natural language processing (NLP) algorithms enable robots to recognize speech and engage in conversational interactions.

Machine learning also plays a crucial role in reinforcement learning, allowing robots to learn from trial and error. Robots can optimize their actions by receiving rewards or penalties in various applications, such as robotics research, gaming, and optimization problems.

In summary, machine learning for robots is the bridge between data and intelligent decision-making, transforming machines into adaptable learners capable of performing complex tasks. Like nurturing a student's growth, machine learning equips robots with the ability to navigate and excel in dynamic and uncertain environments. As technology continues to advance, the synergy between machine learning and robotics holds the promise of creating robots that are not just tools but intelligent companions in our daily lives.

Imagine a robot that can see and understand the world just like we do, recognizing objects, people, and its surroundings with ease. This remarkable capability is achieved through a branch of machine learning known as supervised learning, and it's the key to enhancing robot perception.

Supervised learning is like teaching a robot to recognize different fruits by showing it labeled images of apples, oranges, and bananas. In this process, we provide the robot with a dataset containing input data (images in this case) and corresponding output labels (the names of the fruits). The robot's task is to learn how to map the input data to the correct output labels, allowing it to identify fruits in new, unseen images.

At the heart of supervised learning are neural networks, which are computational models inspired by the human brain's structure and function. These networks consist of interconnected artificial neurons that process and transform data, much like the neurons in our brains process information. Convolutional Neural Networks (CNNs) are particularly well-suited for tasks involving visual data, making them a popular choice for robot perception.

The training process is akin to teaching a student through practice and feedback. It begins with the collection of a labeled dataset, which serves as the training material for the robot. This dataset contains a large number of input-output pairs, allowing the robot to learn the relationships between the inputs and their corresponding outputs.

During training, the robot's neural network adjusts its internal parameters to minimize a loss function, which quantifies how well the network is performing. The goal is to reduce the difference between the predicted outputs and the true outputs in the training data. This iterative process of adjusting parameters continues until the model achieves satisfactory performance.

Validation and testing are crucial steps in supervised learning, like taking quizzes and exams to assess a student's understanding. After training, the model is evaluated on new and unseen data to ensure that it generalizes well beyond the training set. This helps us determine how well the robot can perform real-world tasks.

Overfitting is a common challenge in supervised learning, much like a student who memorizes answers but doesn't truly understand the material. It occurs when a model becomes overly complex and fits the training data too closely, leading to poor performance on new data. Techniques such as regularization and cross-validation are employed to prevent overfitting.

In robot perception, supervised learning is instrumental in tasks like object recognition and image segmentation. For example, a robot equipped with cameras can use supervised learning to identify and classify objects in its environment. It can learn to distinguish between various objects, such as cups, books, and laptops, enabling it to interact intelligently with its surroundings.

One of the remarkable aspects of supervised learning is transfer learning, which is akin to a student applying knowledge from one subject to excel in another. With transfer learning, a pre-trained neural network can be adapted for a new task with limited data. For instance, a neural network initially trained to recognize everyday objects can be fine-tuned to recognize specific medical instruments in a healthcare setting.

The benefits of supervised learning extend to applications beyond object recognition. In autonomous vehicles, such as self-driving cars, supervised learning enables the perception system to identify road signs, pedestrians, and other vehicles.

This knowledge is crucial for safe and efficient navigation on the road.

In the healthcare industry, robots equipped with supervised learning capabilities can assist in medical imaging analysis. They can analyze X-rays, MRIs, and CT scans to detect abnormalities, tumors, or fractures. This not only enhances the speed of diagnosis but also reduces the risk of human error.

In manufacturing, robots use supervised learning for quality control and defect detection. They can inspect products on assembly lines, identify defects or irregularities, and make decisions about whether a product meets quality standards. This ensures that only high-quality products reach consumers.

Supervised learning is also vital for robotic assistants that interact with humans in various contexts. These robots can understand and respond to natural language commands, thanks to NLP (Natural Language Processing) models trained through supervised learning. They can engage in conversations, answer questions, and perform tasks based on verbal instructions.

In summary, supervised learning empowers robots with the ability to perceive and understand the world around them. It allows them to recognize objects, people, and environments, making them more intelligent and capable in a wide range of applications. As technology continues to advance, the synergy between supervised learning and robotics holds the promise of creating robots that are not just tools but intelligent companions in our daily lives.

Chapter 9: Robotics in Real-World Applications

Picture a bustling factory floor where robots tirelessly assemble products, weld components, and perform intricate tasks with precision. This harmonious dance of automation is a testament to the transformative power of industrial robotics and automation.

Industrial robotics, in essence, is the marriage of technology and machinery to create robots specifically designed for manufacturing and industrial processes. These robots are the workhorses of the manufacturing world, capable of performing a wide range of tasks, from repetitive and mundane to highly complex and precise.

Automation, on the other hand, is the overarching concept that encompasses the use of technology to control and monitor processes and machinery with minimal human intervention. It's like having a conductor orchestrating a symphony of machines, ensuring that they perform their tasks seamlessly and efficiently.

The story of industrial robotics and automation begins with the desire to streamline production, increase efficiency, and improve product quality. Early industrial robots were rudimentary compared to their modern counterparts, but they marked the dawn of a new era in manufacturing.

Industrial robots come in various forms, each tailored to specific tasks and industries. Robotic arms, resembling mechanical limbs, are the most common type. These arms are highly flexible and equipped with joints and end-effectors (tools) that enable them to perform tasks like welding, painting, and material handling.

SCARA (Selective Compliance Assembly Robot Arm) robots, akin to skilled artisans, are known for their speed and precision in tasks that require dexterity and accuracy, such as assembling electronics or packaging items.

Delta robots, like nimble acrobats, excel in high-speed pick-and-place operations. They're often used in applications like sorting, packaging, and food processing, where rapid and precise movements are essential.

Mobile robots, much like agile couriers, are designed to navigate and transport materials within industrial environments. They can autonomously move goods from one part of the factory to another, optimizing logistics and reducing manual labor.

Cobots, or collaborative robots, are like friendly coworkers on the factory floor. These robots are equipped with sensors and safety features that allow them to work alongside humans, performing tasks collaboratively. They're particularly valuable in tasks that require human-robot cooperation, such as assembly or quality inspection.

End-effectors, or robot tools, are the equivalent of a craftsman's toolkit. These specialized attachments, like grippers, welding torches, and drills, enable robots to perform a wide range of tasks. The choice of end-effector depends on the specific application and requirements.

Machine vision, akin to robotic eyes, is a crucial component of industrial automation. It allows robots to "see" and make decisions based on visual information. Machine vision systems use cameras and image processing algorithms to inspect and identify objects, ensuring quality control in manufacturing.

Sensors are like the senses of industrial robots, providing them with information about their surroundings. These sensors can detect objects, measure distances, sense temperature, and even capture data about the robot's own movements. Sensors play a pivotal role in enabling robots to operate safely and effectively.

Control systems, like the conductor's baton, are at the heart of industrial robotics. These systems manage and coordinate the robot's movements and actions. They receive input from sensors, process information, and generate commands to control the robot's actuators (motors and joints).

Programming industrial robots is the equivalent of composing a symphony. It involves creating sequences of instructions that dictate the robot's actions. Robot programming languages, such as RAPID and VPL (Visual Programming Language), provide a means of communicating with the robot's control system.

Automation software, similar to a conductor's score, orchestrates the entire manufacturing process. This software coordinates the actions of robots, conveyors, and other machinery to ensure a seamless production flow. It also allows for real-time monitoring and adjustments to optimize efficiency.

Robotic automation cells, like specialized production teams, are integrated systems that combine robots, machinery, and automation software to perform specific tasks. These cells are designed for efficiency and are often used in manufacturing environments to achieve high throughput and quality.

The impact of industrial robotics and automation extends to various industries, revolutionizing the way products are made and processes are carried out. In automotive manufacturing, robots are instrumental in tasks like welding, painting, and assembly, leading to increased precision and production speed.

In the electronics industry, robots handle delicate components with precision and consistency. They can place tiny microchips onto circuit boards or conduct intricate soldering tasks, ensuring the reliability of electronic devices.

Food and beverage processing benefit from automation in tasks like packaging and quality control. Robots can handle food products gently and hygienically, reducing the risk of contamination and improving efficiency.

Pharmaceutical manufacturing relies on automation for tasks like drug formulation and packaging. Robots can precisely dispense ingredients, fill vials, and label products, maintaining strict quality standards.

In logistics and warehousing, robots are like diligent warehouse workers. Autonomous mobile robots can navigate storage facilities, retrieve items, and transport them to fulfillment centers, streamlining order fulfillment processes.

The adoption of industrial robotics and automation continues to grow, driven by the need for increased productivity, cost efficiency, and improved product quality. However, this transformation also raises questions about the future of work and the role of human workers in increasingly automated industries.

In response, there is a growing emphasis on reskilling and upskilling the workforce to adapt to the changing landscape. Rather than replacing human workers, industrial robots and automation are seen as tools that can augment and enhance human capabilities, making work safer and more efficient.

In summary, industrial robotics and automation are the driving forces behind the modern manufacturing landscape. They have transformed factories into highly efficient and flexible production hubs, enabling the creation of complex products with precision and speed. As technology continues to advance, the synergy between robots and humans holds the promise of a harmonious future where automation enhances our capabilities and enriches our lives.

Imagine a world where robots collaborate with healthcare professionals, enhancing patient care, performing delicate surgeries, and automating routine tasks with precision and efficiency. This vision of the future is becoming increasingly real as robots find their place in healthcare and medicine.

Robots in healthcare are like trusted assistants, capable of taking on a wide range of roles to support medical professionals and improve patient outcomes. They are designed to work alongside humans, offering a helping hand in diagnosis, treatment, and patient care.

One of the remarkable applications of robots in healthcare is in surgical procedures. Surgical robots, like skilled surgeons' hands, are equipped with precision instruments and cameras that provide a magnified, 3D view of the surgical site. These robots assist surgeons in performing minimally invasive surgeries, reducing patient trauma, and speeding up recovery times.

In addition to assisting with surgeries, robots can also perform certain medical procedures autonomously. For example, robots can carry out tasks like drawing blood, administering injections, or placing intravenous (IV) lines with high precision. These automation processes reduce the risk of human error and enhance patient safety.

Robotic exoskeletons, like wearable support systems, can assist individuals with mobility impairments. These wearable devices provide mechanical support to help patients walk or perform daily activities. They have the potential to significantly improve the quality of life for people with neurological conditions or spinal cord injuries.

Robots play a crucial role in telemedicine, connecting patients and healthcare providers across distances. Telemedicine robots are equipped with cameras and screens that allow doctors to remotely examine patients, provide consultations, and even prescribe treatments. These robots bridge geographical gaps and ensure that patients receive timely care.

In pharmacy settings, robots automate medication dispensing and management. They can accurately fill prescriptions, sort medications, and package them for distribution. This not only reduces the risk of medication errors but also enhances the efficiency of healthcare facilities.

Robotic devices are employed in physical therapy and rehabilitation, assisting patients in regaining strength and mobility after injuries or surgeries. These devices can provide customized exercises and feedback, allowing therapists to monitor progress and adjust treatment plans accordingly.

Robotic companions, like friendly visitors, offer emotional support and companionship to patients, especially those in long-term care facilities or who are isolated due to health conditions. These robots can engage in conversation, play games, and provide entertainment, reducing feelings of loneliness and improving mental well-being.

The use of artificial intelligence (AI) and machine learning in healthcare is closely intertwined with robotics. AI-powered algorithms can analyze vast amounts of medical data, assisting in diagnosis, predicting disease outcomes, and recommending personalized treatment plans. Robots equipped with AI can provide real-time monitoring of patient vital signs and alert healthcare providers to potential issues.

In research and development, robots are indispensable in drug discovery and laboratory automation. Automated robotic systems can conduct high-throughput screening of potential drug compounds, significantly accelerating the drug development process. Robots also play a role in automating routine laboratory tasks, freeing up scientists to focus on more complex research.

The integration of robotics in healthcare raises important considerations regarding ethics, privacy, and the human-machine relationship. Ensuring the safety and security of patient data, as well as maintaining patient consent and privacy, are paramount concerns.

As robots become increasingly sophisticated and capable, there is a need for standardized regulations and guidelines to govern their use in healthcare. These regulations should address issues such as liability, accountability, and the ethical use of AI in medical decision-making.

The collaboration between robots and healthcare professionals is not about replacing human expertise but rather augmenting it. Robots can handle repetitive and time-consuming tasks, allowing healthcare providers to spend more time with patients, make informed decisions, and provide compassionate care.

In summary, robots in healthcare and medicine represent a promising frontier in the quest for improved patient care and medical advancements. These robots, whether assisting in surgeries, automating pharmacy processes, or providing companionship to patients, have the potential to enhance the efficiency and effectiveness of healthcare systems. As technology continues to advance, the synergy between robots and healthcare professionals holds the promise of a healthier and more accessible future for all.

Chapter 10: Future Trends in Robotics Research

Imagine a world where robots are an integral part of our daily lives, assisting us in numerous tasks and making decisions that impact our well-being. This future is rapidly approaching, and with it come important ethical and social considerations that deserve our thoughtful attention.

One of the foremost ethical concerns in robotics is the question of responsibility. When a robot makes a decision or carries out an action, who should be held accountable if something goes wrong? Is it the robot itself, the human who programmed it, the manufacturer, or the end-user? Determining responsibility is complex, and addressing this issue is crucial for legal and ethical reasons.

Privacy is another significant ethical consideration. As robots become more capable and autonomous, they may have access to sensitive information about individuals and their daily lives. Protecting privacy in the age of robotics involves establishing robust data security measures and clear guidelines on data collection, storage, and usage.

The impact of robotics on employment and the workforce is a topic of ongoing debate. While robots can increase productivity and efficiency, they can also lead to job displacement in certain industries. Preparing for this shift requires proactive measures, such as reskilling and upskilling the workforce to adapt to new roles and industries.

Autonomous robots, capable of making decisions independently, raise questions about moral decision-making. For instance, in a medical setting, if a robot surgeon faces a situation where it must prioritize one patient over another, how should it make that choice? These ethical dilemmas require careful consideration and the establishment of ethical frameworks.

Robot-human interactions introduce a unique set of social considerations. For instance, when robots are designed to resemble humans or animals, people may develop emotional attachments to them. Ensuring that these emotional bonds do not lead to unrealistic expectations or exploitation is a challenge for designers and policymakers.

Bias in AI algorithms used in robotics is a significant ethical concern. If AI systems are trained on biased data, they can perpetuate and amplify existing biases, leading to unfair or discriminatory outcomes. Efforts to mitigate bias and ensure fairness in AI must be a priority.

In healthcare, the use of robots raises complex ethical issues. While robots can enhance patient care and assist in surgeries, they also raise questions about the quality of care and the potential for depersonalization in healthcare interactions. Striking a balance between automation and human touch is essential.

Robot autonomy and decision-making are central ethical concerns in military applications. The development of autonomous weapon systems, capable of making life-or-death decisions, has prompted international discussions on the ethics and regulation of such technology to prevent unintended consequences and violations of international law.

Robots in education bring up ethical issues related to student privacy, data security, and the role of human teachers. While robots can assist in personalized learning and support students with special needs, they must do so while respecting ethical principles and protecting children's data.

The concept of robot rights is a topic that has sparked debates. Some argue that as robots become increasingly sophisticated and autonomous, they may deserve certain rights and protections. Defining the scope of these rights and their implications is a complex and evolving discussion.

Ethical considerations in robotics extend to the treatment of robots themselves. As robots become more lifelike and capable of expressing emotions, questions arise about how humans should treat them. Should there be guidelines or standards for the ethical treatment of robots, even if they lack true consciousness?

Public perception and acceptance of robots also play a role in shaping ethical discussions. Building trust and transparency in robotic systems is essential for their successful integration into society. Public engagement and awareness of the ethical issues surrounding robotics can help guide responsible development and use.

In addressing these ethical and social implications, collaboration between policymakers, researchers, industry leaders, and the public is essential. Ethical frameworks, guidelines, and regulations must be developed to ensure that robotics technology aligns with societal values and priorities.

The field of roboethics has emerged to provide a structured approach to addressing ethical challenges in robotics. Roboethics encompasses a range of principles and guidelines for responsible robot development and use. It encourages multidisciplinary dialogue and cooperation among experts in robotics, ethics, law, and philosophy.

Ultimately, the ethical and social implications of robotics are intertwined with our values and vision for the future. As we navigate the path toward a world where robots are increasingly present, it is our collective responsibility to ensure that these technologies are developed and used in ways that benefit humanity, uphold our ethical standards, and respect our shared values.

Imagine a future where AI and robotics are seamlessly integrated into our daily lives, enhancing productivity, solving complex problems, and improving our quality of life - this future is not a distant dream but a rapidly approaching reality.

AI, or artificial intelligence, is the driving force behind many of the technological advancements we see today. It refers to the development of computer systems that can perform tasks that typically require human intelligence, such as learning, reasoning, problem-solving, and decision-making. Robotics, on the other hand, involves the design and creation of machines capable of physical tasks and interactions with their environment.

The synergy between AI and robotics holds immense potential to reshape industries, transform healthcare, revolutionize transportation, and empower individuals and businesses. In this journey toward an AI-powered future, several key roles emerge that AI and robotics will play:

Automation and Labor Augmentation: One of the most immediate and transformative roles of AI and robotics is automation. Tasks that were once manual and repetitive can now be handled by machines, freeing up human workers to focus on more creative and complex aspects of their jobs. In manufacturing, for example, robots have become integral in assembly lines, improving efficiency and precision.

Enhanced Decision-Making: AI systems have the ability to analyze vast amounts of data in real-time, enabling faster and more informed decision-making across various industries. In finance, AI algorithms can process market data and execute trades with unparalleled speed and accuracy. In healthcare, AI assists in diagnosing diseases and recommending treatment plans based on patient data.

Personalization and Customer Experience: AI-driven personalization is becoming the norm in customer service and marketing. Chatbots and virtual assistants use natural language processing to interact with customers, answer queries, and provide tailored recommendations. This enhances the customer experience and builds brand loyalty.

Healthcare Advancements: AI and robotics are poised to revolutionize healthcare by assisting in diagnostics, surgery, and patient care. Robots can perform surgeries with greater precision, and AI algorithms can analyze medical images to detect diseases at earlier stages. Telemedicine, enabled by AI-powered systems, connects patients with healthcare providers remotely, improving access to care.

Transportation and Mobility: The future of transportation is being reshaped by autonomous vehicles. Self-driving cars and trucks equipped with AI-powered systems have the potential to reduce accidents, ease traffic congestion, and provide greater mobility for individuals with disabilities or limited access to transportation.

Environmental Sustainability: AI and robotics are playing a vital role in addressing environmental challenges. Autonomous drones and robots are used for tasks such as monitoring wildlife, assessing environmental damage, and collecting data on climate change. AI-driven systems can optimize energy consumption in buildings and industries, contributing to sustainability efforts.

Education and Skill Development: AI-powered educational tools and platforms are transforming how we learn and acquire new skills. Personalized learning systems adapt to individual student needs, providing targeted instruction and feedback. Virtual reality (VR) and augmented reality (AR) enhance immersive learning experiences.

Research and Innovation: AI accelerates scientific research and innovation by processing and analyzing vast datasets. Researchers in fields like drug discovery, materials science, and genomics use AI algorithms to identify patterns and make breakthroughs. Robots are employed in laboratories to automate experiments and data collection.

Social and Ethical Considerations: As AI and robotics become more deeply integrated into society, ethical considerations become paramount. Ensuring transparency, fairness, and accountability in AI decision-making processes is essential. Additionally, addressing issues like job displacement and the impact on privacy and security is crucial for responsible development.

Human-Machine Collaboration: The future is not about humans versus machines but rather humans collaborating with machines. The concept of "cobots" (collaborative robots) exemplifies this partnership, where humans and robots work side by side, each contributing their unique strengths.

As we embrace the era of AI and robotics, it's essential to foster collaboration and dialogue among technologists, policymakers, ethicists, and society at large. The responsible development and deployment of AI and robotics require a shared commitment to ethical principles, privacy protection, and ensuring that these technologies serve the greater good.

In summary, the role of AI and robotics in the future is transformative and multifaceted. These technologies hold the promise of improving efficiency, enhancing decision-making, and addressing complex challenges across various domains. While we navigate this evolving landscape, it's essential to harness the potential of AI and robotics while safeguarding our values and ensuring that technology serves humanity's best interests.

BOOK 2
FUNDAMENTALS OF ROBOTICS RESEARCH
BUILDING A STRONG FOUNDATION

ROB BOTWRIGHT

Chapter 1: The Evolution of Robotics: A Historical Perspective

Let's embark on a fascinating journey through time to explore the remarkable world of ancient automata and early mechanical devices. Imagine stepping into the distant past, a world where ingenious inventors and craftsmen crafted mechanical wonders that would leave a lasting mark on the history of technology.

Our journey begins in ancient Greece, where inventors like Ctesibius and Hero of Alexandria laid the foundation for the world of automata. These ingenious minds created devices powered by water, air, and steam. Hero's aeolipile, a simple but ingenious steam engine, captivated imaginations and hinted at the potential of mechanical power.

The Antikythera Mechanism, a true marvel of ancient engineering, takes us to ancient Greece once more. Discovered in a shipwreck off the coast of Antikythera, this sophisticated device, often referred to as the world's first analog computer, was used for astronomical calculations. It demonstrated a deep understanding of complex gear systems and revealed the ancient Greeks' fascination with the cosmos.

Traveling eastward to ancient China, we encounter the incredible craftsmanship of Zhang Heng. In the 2nd century AD, he invented the world's first seismoscope, a device capable of detecting earthquakes and indicating their direction. This early mechanical marvel demonstrated not only scientific acumen but also a commitment to improving society's safety.

Moving forward in time to medieval Europe, we find the exquisite clocks and automata created by skilled artisans. The astronomical clock at the Strasbourg Cathedral, built in the 14th century, not only told time but also displayed the positions of the moon and stars. These mechanical marvels were not just functional; they were also works of art that showcased the creativity and craftsmanship of the time.

In the Islamic Golden Age, inventors like Al-Jazari produced remarkable automata and mechanical devices. Al-Jazari's "Book of Knowledge of Ingenious Mechanical Devices" is a treasure trove of inventions, including programmable humanoid robots that played music. These devices demonstrated a deep understanding of mechanics and were designed to entertain, educate, and serve practical purposes.

Ancient Egypt, known for its architectural wonders, also had its share of mechanical ingenuity. The water clock, or clepsydra, was used to measure time by the steady flow of water from one container to another. This ancient timekeeping device, while relatively simple, played a crucial role in regulating daily life along the Nile.

In the Byzantine Empire, the automata known as "self-striking" or "self-playing" organs added musical charm to palaces and churches. These mechanical organs, operated by a system of levers and weights, delighted audiences with their melodic tunes and were a testament to the craftsmanship of the time.

The Renaissance period ushered in a resurgence of interest in mechanical devices and automata. Leonardo da Vinci, the polymath of the era, sketched designs for various ingenious machines, including mechanical knights, robots, and flying devices. While many of his concepts remained on paper, they inspired future generations of inventors and engineers.

The Enlightenment era brought about the development of mechanical theaters, where intricate scenes and figures were brought to life through a complex system of gears and levers. These mechanical wonders provided entertainment and education, showcasing the beauty of automation.

As we journey through history, it becomes clear that the fascination with automata and early mechanical devices transcended time and culture. These inventions, ranging from the practical to the whimsical, not only showcased human ingenuity but also paved the way for the technological advancements of the modern era.

The legacy of ancient automata and early mechanical devices lives on in today's world of robotics and automation. These humble beginnings laid the groundwork for the sophisticated machines and artificial intelligence that define our present and shape our future.

In summary, the world of ancient automata and early mechanical devices is a testament to human curiosity, creativity, and innovation. These ingenious inventions, born from different corners of the world and different eras of history, continue to inspire and remind us of the enduring quest to understand and harness the power of machines.

Let's delve into the captivating stories of influential figures who have left an indelible mark on the history of robotics. These visionaries, inventors, and innovators have shaped the trajectory of robotics, transforming it from a realm of imagination into a tangible reality that permeates our daily lives.

Our journey begins with George Devol, a name synonymous with the birth of industrial robotics. In 1954, Devol and his partner Joseph Engelberger co-invented the first digitally operated and programmable robotic arm, known as the Unimate. This groundbreaking creation revolutionized manufacturing processes, marking the advent of the modern era of robotics in industry.

Moving to Japan, we encounter the remarkable figure of Shigeo Hirose. A pioneer in the field of biomimetic robotics, Hirose's work was inspired by the elegance and efficiency of natural forms. His innovative snake-like robots, capable of navigating complex terrain and confined spaces, have found applications in search and rescue missions, exploration, and industrial inspection.

In the world of artificial intelligence and robotics, the name Alan Turing stands as a beacon of innovation. Turing's groundbreaking work in the mid-20th century laid the foundation for computer science and AI. His concept of the

Turing machine and his contributions to theoretical computer science continue to influence the development of intelligent machines.

The journey takes us to the realm of science fiction, where Isaac Asimov, a prolific writer and biochemist, envisioned a future where robots coexist with humans. Asimov's Three Laws of Robotics, introduced in his stories, laid the ethical framework for human-robot interactions and influenced discussions on AI ethics and safety.

The pioneering spirit of Joseph F. Engelberger, often referred to as the "Father of Robotics," deserves special mention. Engelberger not only co-invented the Unimate but also played a pivotal role in advocating for the widespread adoption of robotics in industry. His tireless efforts helped establish the field of industrial robotics and its applications in manufacturing.

Jumping to the world of academia, we encounter the brilliant mind of Rodney Brooks. A professor and robotics entrepreneur, Brooks co-founded iRobot, the company behind the popular Roomba vacuum cleaner, and Rethink Robotics, known for its collaborative robots. His research on behavior-based robotics inspired a new paradigm of robotic control, emphasizing simplicity and adaptability.

The path leads us to the captivating story of Cynthia Breazeal, a trailblazer in the field of social robotics. As the creator of Kismet, one of the world's first sociable robots, Breazeal pushed the boundaries of human-robot interaction. Her work laid the groundwork for emotionally expressive robots that can engage with humans on a personal level.

In the realm of humanoid robotics, Hiroshi Ishiguro's work stands out. Ishiguro, a Japanese roboticist, is known for creating androids that closely resemble humans, blurring the line between machine and human. His lifelike robots, such as Geminoid and Telenoid, have applications in fields like telecommunication and human-robot interaction research.

The journey wouldn't be complete without mentioning the contributions of Hans Moravec, a visionary roboticist and AI researcher. Moravec's work in computer vision and mobile robotics paved the way for autonomous navigation and robotic perception. His ideas continue to shape the development of self-driving cars and robotics in agriculture.

In the realm of space exploration, we encounter the legendary figure of Wernher von Braun. Although primarily known for his work on rocketry and the development of the Saturn V rocket, von Braun's contributions extended to robotic systems used in space exploration. His work laid the foundation for robotic missions to other celestial bodies.

Returning to Japan, we find the legacy of Masahiro Mori, known for his concept of the "uncanny valley." Mori's insights into the human response to humanoid robots and lifelike androids have had a profound impact on the design and acceptance of robots in society. His observations continue to influence robot design to evoke positive and comfortable interactions.

In the realm of healthcare robotics, the work of Helen Greiner is truly inspirational. As the co-founder of iRobot and the founder of CyPhy Works, Greiner has pioneered the development of robots for various applications, including telemedicine, search and rescue, and defense. Her dedication to creating robots that improve lives has left a lasting legacy.

These influential figures, each with their unique contributions, have shaped the diverse landscape of robotics. From industrial automation to social robots and beyond, their vision, creativity, and dedication have propelled the field forward. As we continue to advance in the age of robotics, their legacy serves as a reminder of the boundless potential and ethical considerations that accompany the rise of intelligent machines.

Chapter 2: Mathematics and Physics Essentials for Robotics

Welcome to the world of robotics and the role that linear algebra plays in shaping the foundation of this exciting field. Linear algebra, while it might sound complex, is a powerful mathematical tool that forms the backbone of many robotic systems, enabling them to perceive, reason, and act in the world.

At its core, linear algebra deals with vectors and matrices. Now, don't let those terms intimidate you. Think of vectors as arrows in space, representing quantities that have both magnitude and direction. These quantities could be anything from the position of a robot's end effector to the direction of a sensor's reading.

Matrices, on the other hand, are collections of numbers arranged in rows and columns. They're like tables of data that can be used to perform various operations. In robotics, matrices are often employed to represent transformations, which describe how a robot's position or orientation changes as it moves and interacts with its environment.

One of the fundamental concepts in linear algebra is the dot product, which allows us to find the angle between two vectors. This concept is invaluable in robotics because it helps us understand the relative orientation of objects or sensors in space. Imagine a robot arm trying to reach a specific point; knowing the angles between its segments is crucial for accurate positioning.

Now, let's talk about matrices in a robot's world. Transformation matrices are like magical spells that allow us to translate, rotate, and scale objects in a three-dimensional space. Think of them as the instructions that a robot needs to follow to move from one position to another. These transformations are vital for tasks like path planning and robot control.

Speaking of control, linear algebra plays a crucial role in controlling the motion of robots. Controllers use matrices to calculate the forces and torques needed to drive a robot's motors and actuators. This ensures that the robot moves precisely as intended, whether it's picking up objects or performing delicate surgical procedures.

Eigenvalues and eigenvectors might sound like the names of characters from a fantasy novel, but they are essential concepts in linear algebra. They help us understand how a matrix transforms vectors. In robotics, eigenvalues and eigenvectors are used to analyze stability and performance, ensuring that robots behave predictably and safely.

Let's not forget about determinants – they're like the heartbeats of matrices. Determinants tell us whether a matrix has an inverse (meaning it can be reversed), which is crucial for solving equations in robotics. Inverse matrices are used to find the joint angles that will position a robot's end effector precisely where it needs to be.

Linear algebra also plays a vital role in sensor fusion, a process where data from multiple sensors are combined to create a comprehensive understanding of a robot's environment. For example, imagine a self-driving car using data from cameras, lidar, and radar sensors. Linear algebra helps merge this information to make informed decisions, like avoiding obstacles or staying within lane boundaries.

Now, let's talk about linear transformations – these are like the chameleons of robotics. They can change the basis of a vector space, making it easier to work with certain types of data. For example, you might transform data from a camera's perspective to a robot's perspective, simplifying tasks like object recognition.

The concept of rank might remind you of a leaderboard in a video game, but in linear algebra, it tells us something different. The rank of a matrix indicates how many linearly independent columns it has. In robotics, understanding the rank of a matrix is essential for solving equations related to inverse kinematics – determining the joint angles needed to achieve a desired end effector position.

Now, let's shift our focus to singular value decomposition (SVD). It might sound complex, but it's like the Swiss army knife of linear algebra. SVD can be used to analyze the geometry of a robot's workspace, identify redundancies in robot designs, and even improve the accuracy of sensor measurements. It's a versatile tool in the roboticist's toolkit.

You might be wondering about the connection between linear algebra and computer graphics. Well, it's quite strong. Linear transformations are used extensively in computer graphics to create realistic 3D environments and animations. Think about how video games bring virtual worlds to life or how computer-generated imagery (CGI) enhances movies – all thanks to linear algebra.

Let's not forget about the essential concept of least squares optimization. This is like finding the best-fit line through a scatterplot of data points. In robotics, least squares optimization is used in tasks like calibrating sensors, estimating robot poses, and solving inverse kinematics problems. It helps us find the most accurate solutions in noisy and uncertain environments.

Now, let's journey into the realm of robot vision. Linear algebra is the backbone of computer vision, enabling robots to interpret images and make sense of their surroundings. Matrices and transformations play a crucial role in tasks like image registration, feature tracking, and object recognition.

Eigenvalue decomposition (EVD) is another powerful tool in linear algebra's arsenal. It helps us break down complex matrices into simpler forms, making it easier to analyze and understand their behavior. In robotics, EVD is used in areas like control theory and stability analysis, ensuring that robots perform reliably and predictably.

Matrix factorization might sound like breaking down a secret code, but in robotics, it's about breaking down complex problems. By factoring matrices into simpler components, roboticists can tackle challenging tasks like simultaneous localization and mapping (SLAM), which enables robots to navigate and map unknown environments.

Now, let's venture into the world of machine learning and deep learning. Linear algebra forms the foundation of many machine learning algorithms, allowing robots to learn from data and make intelligent decisions. Matrix operations are at the heart of neural networks, enabling robots to recognize patterns, make predictions, and even play games.

In summary, linear algebra is the mathematical glue that holds the diverse elements of robotics together. It empowers robots to perceive their environment, make decisions, and interact with the world. Whether it's controlling a robotic arm with precision or enabling a self-driving car to navigate safely, linear algebra is the silent hero that brings the magic of robotics to life. So, the next time you see a robot in action, remember the role that linear algebra plays in making it all possible – it's the language of robots, translating complex mathematics into real-world actions.

Welcome to the exciting world of robotics, where the laws of physics form the very foundation of design and operation. As we embark on this journey, it's important to understand that robotics is not just about clever programming and mechanical ingenuity; it's about harnessing the principles of physics to create machines that can sense, move, and interact with the world around them.

Let's start with one of the most fundamental principles in physics: Newton's Laws of Motion. These laws, established by Sir Isaac Newton, provide the basis for understanding how objects move and interact with forces. In the context of robotics, these laws are like the guiding principles that dictate how robots navigate their environment and perform tasks.

Newton's First Law, often called the law of inertia, states that an object at rest tends to stay at rest, and an object in motion tends to stay in motion unless acted upon by an external force. This law is at the heart of robot stability and control. When a robot is in motion, it requires precise control and feedback systems to maintain its desired trajectory.

Newton's Second Law introduces us to the concept of force, stating that the force applied to an object is equal to the mass of the object multiplied by its acceleration (F = ma). In robotics, this law plays a pivotal role in calculating the forces needed to move robot limbs, manipulate objects, and navigate various terrains. It's the science behind a robot's ability to lift heavy objects or accelerate and decelerate smoothly.

Newton's Third Law, often summarized as "for every action, there is an equal and opposite reaction," underlies the mechanics of robot locomotion. When a robot moves, its actions generate reactions on its environment, and understanding these reactions is crucial for stability and efficient motion. Think of a walking robot taking a step – the ground exerts an equal and opposite force to keep it balanced.

Now, let's venture into the world of energy and work. The principle of conservation of energy tells us that energy cannot be created or destroyed but can only change forms. In robotics, this principle guides the design of energy-efficient systems. Robots must manage their energy resources, whether it's the power source for a mobile robot or the energy required for the precise movements of a robotic arm.

The concept of torque, a rotational force, is a cornerstone of robot dynamics. Torque is what enables robotic joints to rotate and move. Think of a robot arm picking up an object – the torque applied to its joints determines how effectively it can lift and manipulate that object.

In the realm of electromagnetism, robots often rely on principles such as electromagnetic induction and electromagnetic fields. These principles are employed in sensors, actuators, and power systems. For instance, electromagnetic sensors can detect changes in magnetic fields, allowing robots to perceive their surroundings and navigate with precision.

The study of fluids and fluid dynamics is essential for underwater and aerial robots. Understanding buoyancy, pressure, and flow is critical for the design and operation of submarines, drones, and other vehicles that operate in fluid environments. It's like giving robots the ability to swim through water or glide through the air.

Now, let's delve into the world of optics and light. Robots equipped with cameras and sensors rely on the principles of optics to perceive the world visually. Understanding how light travels, reflects, and refracts enables robots to capture images, recognize objects, and navigate based on visual cues.

Thermodynamics, the study of heat and energy transfer, is crucial for robots operating in extreme environments. Whether it's a robot exploring the depths of the ocean or the harsh conditions of outer space, managing heat and energy is essential for survival and functionality.

The principles of sound and acoustics are valuable for robots that rely on auditory sensors or communication systems. Think of a robot that uses sonar to detect objects underwater or a voice-activated assistant that responds to your commands – these technologies rely on the physics of sound waves.

Quantum mechanics, although not a primary focus in most robotics applications, plays a role in the development of advanced sensors and materials. Quantum sensors can provide ultra-precise measurements for tasks like navigation and environmental monitoring, pushing the boundaries of what robots can achieve.

In the realm of materials science, engineers explore the properties of different materials to create robots with specific capabilities. For example, shape-memory alloys can be used in robotic actuators to provide flexibility and adaptability in a robot's movements.

Lastly, let's not forget about the critical role of feedback and control systems. In robotics, these systems are like the conductor of an orchestra, ensuring that the robot's movements and actions align with its objectives. They use sensors to gather data, apply algorithms to make decisions, and send commands to actuators, all in real-time.

In summary, the principles of physics are the invisible threads that weave through the fabric of robotics. They govern how robots move, sense, and interact with the world, shaping the design and capabilities of these remarkable machines. As we continue to push the boundaries of what robots can do, our understanding of physics will remain an essential guide, helping us navigate the exciting frontier of robotics. So, as you delve deeper into the world of robotics, remember that behind every robot's actions and capabilities lies a profound understanding of the laws that govern our physical universe.

Chapter 3: Robotic Sensors and Perception Systems

Welcome to the fascinating realm of sensor fusion, where the power of multiple sensors converges to provide robots with a richer, more comprehensive perception of their surroundings. In the world of robotics, perception is everything. Robots need to understand their environment, detect obstacles, recognize objects, and make sense of the data they collect to navigate and interact effectively.

Imagine a robot exploring a cluttered room. It needs to know where it is, avoid collisions with objects, identify doors, and locate specific items. Achieving these tasks with a single sensor can be challenging, as sensors have limitations and blind spots. However, when we combine the data from multiple sensors, a new world of possibilities opens up.

Let's begin by understanding the concept of sensor fusion. It's like having a team of experts with different skills collaborating to solve a complex problem. In this case, the experts are sensors, each specialized in a particular aspect of perception – vision, lidar, radar, ultrasonics, and more. By bringing their collective insights together, robots can create a more accurate and reliable picture of their environment.

Consider the power of visual sensors, such as cameras. Cameras provide detailed information about the colors, shapes, and textures of objects. They excel at tasks like object recognition and tracking. However, they have limitations in low-light conditions or when objects are occluded from view.

Now, introduce lidar sensors into the mix. Lidar emits laser beams to measure distances and create 3D maps of the surroundings. Lidar is excellent at detecting obstacles, precisely measuring distances, and providing a robust perception of the environment. However, it might struggle to identify objects based on appearance alone.

Radar sensors, on the other hand, use radio waves to detect objects and their velocities. They are particularly valuable in adverse weather conditions, where other sensors might falter. Radar is like the robot's sixth sense, enabling it to perceive the world even in challenging situations.

Ultrasonic sensors operate on a different principle, emitting sound waves and measuring the time it takes for them to bounce back. These sensors are perfect for close-range detection and are often used in robotics for tasks like object avoidance and parking assistance. Magnetic sensors, while less common, can be used for navigation and orientation in certain environments. They detect changes in the Earth's magnetic field, allowing robots to determine their position and heading.

Now, imagine a robot equipped with a combination of these sensors. It's like giving the robot a suite of tools, each specialized in a specific aspect of perception. When it encounters an obstacle, it can use lidar to precisely measure the distance and shape of the obstacle, radar to assess its speed, and camera to identify if it's a pedestrian or another vehicle.

Sensor fusion involves harmonizing the data from these diverse sensors to create a coherent and accurate perception of the world. This process often starts with sensor calibration, ensuring that all sensors provide data in a common reference frame. Think of it as making sure everyone speaks the same language in the team of experts.

Once the data is collected, sensor fusion algorithms come into play. These algorithms process the sensor data, align it in time and space, and merge it into a unified representation of the environment. It's like the experts collaborating and sharing their findings to create a comprehensive report.

One of the key advantages of sensor fusion is redundancy. By having multiple sensors providing overlapping information, robots can maintain functionality even if one sensor fails or encounters limitations. It's like having backup experts who can step in when needed.

Furthermore, sensor fusion enhances the reliability of perception. When multiple sensors independently detect the same object or obstacle, the robot can be more confident in its understanding of the environment. It's like getting a second opinion from multiple experts.

Consider a self-driving car navigating a complex urban environment. It relies on a combination of sensors, including cameras, lidar, radar, and ultrasonics. Cameras help identify traffic signs and lane markings, lidar creates a detailed map of the surroundings, radar detects nearby vehicles, and ultrasonics help with parking and low-speed maneuvering.

In such scenarios, sensor fusion is not just an option; it's a necessity. The car needs to make split-second decisions to ensure safety and efficiency. By fusing data from various sensors, it can accurately perceive its surroundings, predict the behavior of other road users, and make informed decisions, such as when to brake or change lanes.

Sensor fusion also plays a vital role in robotics beyond autonomous vehicles. It's used in industrial robots for tasks like precise positioning and object manipulation. In agriculture, robots equipped with multiple sensors can navigate fields, detect pests, and optimize crop management. Search and rescue robots benefit from sensor fusion to locate survivors in disaster-stricken areas.

In healthcare, robots with vision and tactile sensors can assist in surgeries, ensuring precise and safe procedures. Drones rely on sensor fusion to navigate complex environments, avoid obstacles, and deliver packages. Even household robots, like vacuum cleaners, use sensor fusion to navigate and clean efficiently.

As technology advances, sensor fusion continues to evolve. Machine learning and artificial intelligence are now integrated into sensor fusion algorithms, enabling robots to adapt and learn from their environments. This adaptive capability allows robots to improve their perception over time, making them more versatile and capable.

In summary, sensor fusion is like giving robots a multi-sensory perception of the world. It empowers them to navigate, interact, and make decisions in complex and dynamic environments. By combining the strengths of different sensors, robots become more versatile, reliable, and capable of performing a wide range of tasks. So, the next time you see a robot in action, remember that behind its seamless perception lies the power of sensor fusion, harmonizing data from multiple sensors to create a unified understanding of the world.

Welcome to the captivating realm of computer vision, where robots gain the ability to see and make sense of the visual world, much like humans do. In the grand mosaic of robotics, computer vision is a crucial piece, enabling robots to perceive, understand, and interact with their surroundings using cameras and sophisticated algorithms.

Think of computer vision as the eyes of a robot. Just as our eyes capture and process visual information to navigate, recognize objects, and make decisions, computer vision equips robots with a similar capability, albeit in a digital and computational form.

At its core, computer vision involves the use of cameras to capture images or video streams from the robot's perspective. These images serve as the raw data from which the robot will derive meaning. Imagine a robot equipped with a camera, observing its environment much like you would observe a room through a pair of glasses.

Once the robot has gathered visual data, the real magic begins with image processing and analysis. Think of this stage as the robot's brain interpreting the images it sees. It's like a detective examining clues to solve a mystery. The robot's algorithms work tirelessly to identify patterns, shapes, colors, and objects within the images.

One of the fundamental tasks in computer vision is image segmentation. This process involves dividing an image into meaningful regions or segments. It's like drawing lines around objects in a picture, separating them from the background. For example, in a room, segmentation helps the robot identify furniture, people, and other objects.

Object recognition is another pivotal aspect of computer vision. It's like the robot's ability to name and categorize the objects it sees. By comparing the features of objects in the images to a database of known objects, the robot can recognize and label them. Think of it as the robot saying, "That's a chair, that's a table, and that's a person."

Now, let's dive into the fascinating world of feature extraction. Features are distinctive characteristics of objects, such as edges, corners, or textures. Imagine a robot scanning an image and marking key points that represent these features. These points serve as landmarks that help the robot identify and match objects.

Machine learning plays a significant role in computer vision. Just as we learn to recognize objects through experience, robots can be trained to do the same. Machine learning algorithms analyze vast datasets of images, learning to recognize patterns and objects. This process allows robots to improve their accuracy and adapt to new environments.

Consider the example of a robotic vacuum cleaner. Equipped with a camera and computer vision, it scans a room to identify obstacles like furniture or toys. By recognizing these objects, it can plan a path to avoid collisions, much like how we navigate a cluttered room without bumping into things.

Another exciting application is facial recognition. Robots can be programmed to identify and interact with humans based on their facial features. Imagine a robot that welcomes you with a friendly "Hello" when it recognizes your face. It's like having a personalized greeting from a digital friend.

Object tracking is another remarkable capability of computer vision. It's like the robot keeping an eye on moving objects. Whether it's tracking a soccer ball during a game or following a person walking through a crowd, object tracking enables robots to interact with dynamic environments.

Depth perception is crucial for robots to understand the 3D structure of their surroundings. Just as our two eyes provide depth cues through binocular vision, robots can use stereo cameras or depth sensors to calculate distances and create 3D representations of the environment. This capability is essential for tasks like navigating through complex terrain.

Semantic segmentation takes computer vision a step further by not only identifying objects but also understanding their context. Imagine a robot that not only recognizes a traffic sign but also understands its meaning and obeys traffic rules. It's like having a robot that comprehends the world around it.

Object manipulation is an exciting application of computer vision in robotics. Robots equipped with cameras and sophisticated algorithms can pick up and manipulate objects with precision. Think of a robot assembling intricate components or sorting items on a conveyor belt – all guided by its visual perception.

Visual SLAM, or Simultaneous Localization and Mapping, is like a robot creating a map of its surroundings while simultaneously determining its own position within that map. It's a crucial capability for autonomous robots, allowing them to navigate unknown environments and build a mental map of their surroundings as they go.

Autonomous vehicles, such as self-driving cars, rely heavily on computer vision. They use cameras, lidar, radar, and other sensors to perceive the road and surrounding vehicles, pedestrians, and traffic signs. Computer vision algorithms help these vehicles make real-time decisions for safe and efficient navigation.

In summary, computer vision is the art and science of giving robots the gift of sight. It's the technology that allows them to perceive, recognize, and interact with the visual world. From recognizing objects to navigating complex environments, computer vision empowers robots to perform a wide range of tasks. So, the next time you see a robot with a camera "looking" at its surroundings, remember that it's not just seeing – it's engaging in the marvelous world of computer vision, deciphering the visual puzzle of our world, one pixel at a time.

Chapter 4: Robotic Actuators and Manipulators

Welcome to the fascinating realm of robotic actuators, the dynamic muscles and joints that bring robots to life with motion and functionality. In the world of robotics, actuators are like the performers in a grand symphony, orchestrating precise movements and enabling robots to interact with the physical world.

Let's start by understanding the fundamental concept of actuation. At its core, actuation is the process of converting energy into motion. Think of it as the driving force that propels a robot forward, moves its arms, and powers its various mechanical components. Without actuators, robots would remain static and lifeless, like statues in a park.

Actuators come in various forms, each with its unique strengths and applications. The most common types of actuators in robotics are electric, pneumatic, hydraulic, and piezoelectric actuators. Each of these types has its advantages and is suited for specific tasks and environments.

Electric actuators, for instance, are the workhorses of robotics. They rely on electric motors to generate motion. Think of them as the engines that power your car, but on a smaller scale. Electric actuators are versatile, precise, and efficient, making them ideal for tasks that require accuracy and control, such as robot arms assembling delicate components.

Pneumatic actuators, on the other hand, use compressed air to create motion. Picture the whooshing sound of an air compressor inflating a tire – that's the power behind pneumatic actuators. They are fast, lightweight, and well-suited for applications that require rapid movements, like industrial robots performing pick-and-place operations.

Hydraulic actuators are like the strongmen of the robot world. They use pressurized fluid, typically oil, to generate powerful and robust motion. Hydraulic actuators are ideal for heavy-duty tasks, such as lifting heavy objects or operating construction equipment. They can exert tremendous force while maintaining precision.

Piezoelectric actuators are the precision artists among actuators. They use the piezoelectric effect, where certain materials change shape when an electric voltage is applied, to create precise, nanoscale movements. These actuators are incredibly accurate and find applications in fields like microscopy and nanopositioning.

One of the critical aspects of actuators is their ability to provide feedback. Think of this feedback as the sense of touch for a robot. Encoders and sensors are often integrated into actuators to measure position, velocity, and force. This feedback allows robots to adjust their movements in real-time, ensuring accuracy and safety.

Now, let's explore the anatomy of a robotic actuator. At the core of most actuators is a mechanism called a rotary or linear drive. Rotary drives convert rotational motion into linear motion and vice versa. Think of a spinning wheel that can move a robot's arm up and down or side to side.

In contrast, linear drives directly produce linear motion, much like a piston in an engine. Linear actuators are often used in applications where straight-line movement is essential, like opening and closing a robotic gripper.

Gears play a crucial role in actuation, much like the gears in a bicycle. They enable robots to trade speed for torque or vice versa. Gearboxes are commonly used to increase the force generated by an actuator while sacrificing some speed.

Imagine a robot arm picking up a heavy object. The actuator near the base of the arm might use a gearbox to provide the necessary torque to lift the object, while the actuator at the arm's tip might prioritize speed and accuracy for delicate tasks.

Solenoids are another type of actuator, often used for on-off control of various mechanical components. Think of them as the digital switches of the robotic world. Solenoids are used in applications like valve control, where precise fluid flow is required.

Muscles and tendons are inspiration for some advanced robotic actuators. Bio-inspired actuators mimic the way muscles contract and relax to produce motion. These actuators offer flexibility and adaptability, making them suitable for applications like prosthetics and biomimetic robots.

Actuators can be found throughout a robot's body, from its limbs to its joints. Take a robot arm, for example. It may have multiple actuators to control each joint, allowing for a wide range of movements. Think of it as a series of hinges and motors working together to mimic the flexibility of a human arm.

One of the key challenges in robotics is designing actuators that are both powerful and energy-efficient. Robots often operate on limited power sources, such as batteries, so maximizing efficiency is crucial. Engineers continuously strive to develop actuators that can deliver high performance while minimizing energy consumption.

Now, let's delve into the concept of actuator control. Just as a conductor guides an orchestra, robot programmers orchestrate the movements of actuators to perform tasks. The control system sends commands to the actuators, specifying how they should move and when.

Control algorithms are the secret sauce behind precise robotic movements. They calculate the desired positions and velocities of actuators based on the robot's objectives and sensor feedback. These algorithms ensure that a robot can paint a picture, perform surgery, or explore distant planets with grace and accuracy.

In collaborative robots, or cobots, safety is a top priority. Specialized actuators equipped with sensors can detect unexpected collisions and stop the robot's movement to prevent harm to humans. It's like a robot having a built-in safety net to protect itself and those around it.

In summary, robotic actuators are the engines of the robot world, breathing life and motion into mechanical bodies. Whether it's the precise movements of a surgical robot, the strength of an industrial robot, or the delicacy of a robotic hand, actuators are the unsung heroes of robotics, making tasks that were once the realm of science fiction a reality. So, the next time you see a robot gracefully performing a task, remember that beneath its mechanical exterior lies the intricate world of actuators, diligently converting energy into purposeful motion, one movement at a time.

Welcome to the captivating world of manipulator arm design and kinematics, where robots are equipped with versatile and dexterous limbs that mimic the movements of the human arm. In the grand symphony of robotics, the manipulator arm is like the virtuoso soloist, capable of performing a wide range of precise and intricate tasks.

To understand manipulator arm design, let's begin with the concept of kinematics. Kinematics is the branch of robotics that focuses on describing the motion of robotic limbs without considering the forces involved. It's like studying the graceful movements of a ballet dancer without thinking about the strength required to perform those movements.

A manipulator arm typically consists of multiple segments or links connected by joints. Imagine a robotic arm resembling a human arm with shoulder, elbow, and wrist joints. These joints are where motion happens, and they allow the arm to articulate and reach different positions.

The number and type of joints in a manipulator arm depend on its intended application and complexity. Some robots have simple arms with a few joints, while others boast highly articulated arms with multiple degrees of freedom. Degrees of freedom refer to the number of independent ways a joint can move, much like the flexibility of your own joints.

Now, imagine a robot arm reaching out to grab an object. The first joint near the base of the arm corresponds to the shoulder joint in a human arm. It allows the arm to move up and down, similar to your shoulder's motion. This joint provides the robot's arm with its first degree of freedom.

The second joint, often located at the elbow of the arm, is like your elbow joint. It enables the arm to bend and extend, providing a second degree of freedom. Just like you can bend your arm to reach for objects, this joint allows the robot's arm to change its orientation.

The wrist joint is like the final piece of the puzzle. It offers the third degree of freedom, allowing the arm to twist or rotate. Think of it as the joint that lets you turn your hand palm-up or palm-down. For a robot, this joint is crucial for orienting its end effector—the tool or gripper at the arm's tip.

The end effector is the business end of the manipulator arm. It's like the robot's hand, and it can take various forms depending on the task at hand. It could be a gripper for picking up objects, a welding torch for joining materials, or a camera for inspection. The design of the end effector is tailored to the specific application of the robot.

Now, let's discuss inverse kinematics. Inverse kinematics is the art of figuring out how to position the joints of the manipulator arm to achieve a desired end effector position and orientation. It's like solving a puzzle where you know the destination but need to find the right moves to get there.

Think of a robot trying to place a book on a shelf. The robot knows where it wants to place the book (the desired end effector position), but it needs to calculate the joint angles to reach that position. Inverse kinematics algorithms do precisely that – they determine the joint angles that achieve a given end effector pose.

Inverse kinematics is crucial for tasks that require precise positioning, like assembling electronic components or performing surgery. It's the mathematical magic that ensures the robot's arm moves with grace and accuracy, much like a skilled dancer performing a choreographed routine.

Now, let's explore the concept of forward kinematics. Forward kinematics is the process of determining the end effector's position and orientation based on the joint angles of the robot's manipulator arm. It's like tracing the path of a dancer's hand in a dance routine.

Imagine a robot with known joint angles for each of its arm's joints. Forward kinematics equations can compute the position and orientation of the end effector. This information is vital for tasks like 3D printing, where the robot needs to precisely position the print head. Inverse and forward kinematics are like two sides of the same coin, working together to enable the robot to plan and execute its movements. While inverse kinematics helps the robot reach its target, forward kinematics allows it to understand where it is in space.

Now, consider the importance of workspace analysis. The workspace is like the playground where the robot can reach and interact with objects. It's the region in space that the manipulator arm can access. Workspace analysis helps engineers and roboticists understand the limitations and capabilities of a robot's arm.

Imagine a robot used for painting large structures. Engineers need to ensure that the robot's arm can reach all the nooks and crannies of the structure's surface. Workspace analysis involves modeling the arm's geometry and calculating the reachable points in space.

Singularity analysis is another critical aspect of manipulator arm design. Singularities are like the moments in a dance routine where the dancer's movements become challenging to control. In robotics, singularities are configurations where the arm loses degrees of freedom, making certain movements difficult or impossible.

Think of a robot arm reaching a point where two of its joints align perfectly. At this singularity, the arm loses a degree of freedom, and it becomes challenging to control its movements.

Singularity analysis helps roboticists identify and avoid these problematic configurations.

In summary, manipulator arm design and kinematics are the choreography and dance of the robotic world. They enable robots to move with precision, grace, and purpose, performing tasks ranging from delicate surgeries to manufacturing and exploration. So, the next time you see a robot's arm gracefully performing a task, remember that behind its movements lies the intricate world of manipulator arm design and kinematics, orchestrating the beauty of motion in the realm of robotics.

Chapter 5: Robot Control and Navigation

Welcome to the captivating world of robot path planning and trajectory generation, where robots navigate through complex environments with the grace and precision of a skilled dancer. In the grand choreography of robotics, path planning and trajectory generation are like the carefully rehearsed steps that ensure a robot moves from point A to point B without missing a beat.

Imagine a robot tasked with delivering a package in a cluttered room. Path planning is the art of finding the best route for the robot to follow, avoiding obstacles, and reaching its destination efficiently. It's like a GPS guiding you through a maze of city streets to your favorite coffee shop.

One of the fundamental aspects of path planning is collision avoidance. Just as you instinctively steer clear of obstacles when walking through a crowded room, path planning algorithms help robots avoid collisions with objects in their environment. These algorithms calculate safe paths by considering the robot's size, shape, and the positions of obstacles.

Now, let's talk about the concept of configuration space. Configuration space is like a map of all possible positions the robot can occupy. Imagine a robot moving in a 2D environment – its configuration space would represent all the x and y coordinates where the robot's center could be located. Path planning algorithms often work in this configuration space to find collision-free paths.

Consider a robot arm with multiple joints. Its configuration space becomes more complex, resembling a multidimensional puzzle. Path planning algorithms navigate this high-dimensional space to determine joint angles and positions that enable the arm to reach its target without collisions.

One of the challenges in path planning is dealing with dynamic environments. Imagine a robot navigating a busy airport terminal filled with people moving in unpredictable ways. Dynamic path planning involves continuously updating the robot's path in real-time to avoid collisions with moving objects.

Trajectory generation is like adding style and finesse to the robot's movements. While path planning finds the best route, trajectory generation determines how the robot should move along that route. It's like choreographing a dance routine for the robot, specifying its speed, acceleration, and smoothness of motion.

Consider a robot arm tasked with painting a mural on a wall. Trajectory generation algorithms dictate how the arm moves its paintbrush – whether it moves smoothly or with deliberate pauses, and how it transitions between different strokes and colors.

Smoothness is a crucial aspect of trajectory generation. Just as a skilled dancer's movements flow seamlessly from one step to the next, smooth trajectories ensure that a robot's motions are graceful and free of jerky movements. This is especially important in applications like robot-assisted surgery, where precise and fluid motions are critical.

Real-time trajectory generation is essential for tasks that require immediate responses. Imagine a robot on a factory floor that needs to adjust its path to avoid a sudden obstacle. Real-time algorithms rapidly generate new trajectories to ensure the robot can navigate around the obstacle without delay.

Now, let's dive into the concept of optimization. Path planning and trajectory generation often involve optimizing certain criteria. For example, an optimization algorithm might aim to find the shortest path, the fastest path, or the path that conserves the most energy.

Imagine a delivery drone optimizing its flight path to deliver packages more efficiently. The optimization algorithm considers factors like wind speed, battery life, and package weight to find the most economical route.

One of the exciting developments in path planning and trajectory generation is the integration of machine learning. Machine learning algorithms can adapt and improve path planning and trajectory generation based on experience. They can learn from past successes and failures to make better decisions.

Consider a self-driving car using machine learning to navigate complex urban environments. Over time, it learns to anticipate traffic patterns, pedestrian behavior, and road conditions, allowing it to plan more effective paths and trajectories.

Cooperative path planning involves multiple robots working together to achieve a common goal. It's like a group of dancers performing a synchronized routine. Cooperative algorithms coordinate the movements of robots to avoid collisions and optimize their paths collectively.

Think of a team of autonomous drones tasked with surveying a large area. Cooperative path planning algorithms ensure that the drones can cover the area efficiently without colliding or duplicating efforts.

In summary, path planning and trajectory generation are the choreographers of the robotic world, guiding robots through intricate dances in complex environments. Whether it's a robot delivering packages, a surgical robot performing delicate procedures, or a self-driving car navigating city streets, these algorithms ensure that robots move with precision and grace. So, the next time you witness a robot gracefully maneuvering through its tasks, remember the artistry of path planning and trajectory generation, orchestrating the elegance of robotic motion in our increasingly automated world.

Welcome to the intriguing world of localization and Simultaneous Localization and Mapping (SLAM), where robots navigate and understand their surroundings, much like explorers mapping uncharted territories. In the realm of robotics, these concepts are like a GPS combined with a cartographer's skills, allowing robots to determine their position and create maps of the world around them.

Localization is the process by which a robot determines its position within a known environment. Imagine you are in a pitch-dark room, and you need to figure out exactly where you are without bumping into furniture or walls. Localization provides the robot with its sense of place.

Now, consider a robot in a familiar room equipped with sensors like cameras and laser scanners. These sensors collect data about the robot's surroundings, such as the position of objects and the distance to walls. Localization algorithms use this sensor data to estimate the robot's precise location within the room.

One common method for localization is known as odometry, which involves tracking the robot's movements over time. Think of it as counting your steps to determine how far you've traveled. Robots equipped with wheel encoders can measure the rotations of their wheels and calculate their movement accurately.

However, odometry alone is not always sufficient for precise localization, especially in complex environments with uncertainties like uneven terrain or wheel slippage. To address these challenges, robots often use a combination of sensors, such as cameras, lidar, and GPS, to improve their localization accuracy.

Simultaneous Localization and Mapping, or SLAM, takes localization to the next level by not only determining the robot's position but also creating a map of its environment as it moves. Imagine you're exploring a new city and, as you walk, you're sketching a map of the streets, landmarks, and buildings around you. That's precisely what SLAM enables robots to do.

SLAM algorithms process sensor data in real-time to build a map of the robot's surroundings while simultaneously estimating the robot's position within that map. This is a bit like a detective piecing together a puzzle, where each new sensor measurement helps refine the map and the robot's location.

Consider a robot exploring an unknown building. As it moves, its laser scanner measures the distances to nearby walls and objects. SLAM algorithms use this data to create a map of the building's layout, including walls, doors, and furniture. At the same time, the robot's position within the map is continuously updated, allowing it to navigate confidently.

One of the key challenges in SLAM is dealing with sensor noise and uncertainty. Sensors can provide inaccurate measurements due to various factors like reflections, occlusions, or sensor limitations. SLAM algorithms must be robust enough to handle this uncertainty and still produce reliable maps and localization estimates.

Another important aspect of SLAM is loop closure detection. Imagine you're walking in a large park and suddenly find yourself back at a familiar spot. You recognize the location because of a distinctive tree or landmark you passed earlier. Loop closure detection is like the robot's ability to recognize when it has revisited a place it has been before. It helps correct errors in the map and localization estimates, ensuring accuracy over time.

Visual SLAM is a variant of SLAM that relies on cameras to capture images of the environment. Think of it as a robot using its "eyes" to create a map and determine its position. Visual SLAM has applications in areas like autonomous drones, where cameras provide rich visual data for mapping and navigation.

Consider a drone flying through a forest, capturing images of the trees and terrain below. Visual SLAM algorithms analyze these images to create a 3D map of the forest and keep track of the drone's position. This enables the drone to fly safely and avoid obstacles.

Mobile robots are not the only ones benefiting from SLAM. Autonomous vehicles, such as self-driving cars, rely on SLAM algorithms to navigate complex urban environments. These vehicles use a combination of lidar, cameras, and other sensors to create maps of the road, detect other vehicles and pedestrians, and ensure safe navigation.

In addition to terrestrial applications, SLAM is also vital in underwater exploration and planetary exploration. Imagine a robot exploring the ocean depths or the surface of Mars. SLAM helps these robots create maps of their alien environments and determine their precise locations, even in the absence of GPS signals.

One of the exciting frontiers in SLAM research is the fusion of multiple sensor modalities. Imagine a robot combining data from cameras, lidar, and radar to create highly detailed maps and achieve precise localization. Sensor fusion allows robots to adapt to various environments and navigate with greater accuracy.

In summary, localization and SLAM are the navigational senses and mapmaking abilities that empower robots to explore and understand the world around them. Whether it's a robot in a factory, a self-driving car on a bustling street, or a rover on an alien planet, these technologies enable robots to find their way and interact with their surroundings with remarkable accuracy and sophistication. So, the next time you see a robot confidently navigating its environment, remember the magic of localization and SLAM, guiding it through the uncharted territory of our ever-evolving robotic landscape.

Chapter 6: Kinematics and Dynamics of Robotic Systems

Welcome to the fascinating realm of forward kinematics and inverse kinematics, two essential concepts that breathe life into the movements of robotic arms and legs, making them dance to the rhythm of precision and purpose.

Forward kinematics is like the choreography that plans the robot's every move on the stage. It's the process of determining the position and orientation of the robot's end effector, the hand or tool at the tip of its limb, based on the angles of its joints. Think of it as orchestrating a ballet where each joint in the robot's limb is a dancer, and the end effector is the lead performer.

Imagine a robotic arm with multiple joints, much like a human arm with its shoulder, elbow, and wrist. In forward kinematics, we map the joint angles to the position and orientation of the end effector. This calculation is like understanding how the angles of your own arm joints create the movement of your hand when you reach for something.

Now, consider a robot arm in a factory assembling products on an assembly line. Forward kinematics allows the robot to precisely position its end effector to pick up components, assemble them, and place the finished product with grace and accuracy.

Inverse kinematics, on the other hand, is the art of figuring out the joint angles needed to achieve a desired position and orientation of the end effector. It's like a puzzle where you know where you want to place a piece, and you need to find the right moves to get it there.

Imagine you're playing with a robotic arm, and you want it to place a toy in a specific spot on a shelf. Inverse kinematics is the mathematical magic that calculates the joint angles required to make the robot's end effector place the toy exactly where you want it.

Inverse kinematics plays a critical role in various applications, from robotic surgery, where precise movements are essential, to animation in the film industry, where characters move in natural and lifelike ways.

Consider a surgeon using a robotic surgical system to perform minimally invasive surgery. The surgeon manipulates a console, and the robotic arms translate those movements into precise actions inside the patient's body. Inverse kinematics ensures that the surgeon's commands are faithfully executed by the robot.

Now, let's dive deeper into the mathematics of forward and inverse kinematics. These calculations involve matrices, trigonometry, and complex equations, much like the movements of a dancer are guided by a choreographer's intricate instructions.

For forward kinematics, we use transformation matrices to represent the relationship between each joint's movement and the end effector's position and orientation. These matrices allow us to perform matrix multiplications to calculate the final pose of the end effector.

Think of it as a dance routine where each dancer's movements are coordinated using a set of instructions. In the world of robotics, these instructions are represented by matrices that transform joint angles into end effector positions.

Inverse kinematics, while mathematically more complex, is akin to solving a puzzle. We use algorithms and mathematical techniques, like the Jacobian matrix, to determine the joint angles that will position the end effector where we desire.

Imagine a jigsaw puzzle where you know the final picture, but you need to figure out how to arrange the pieces. Inverse kinematics algorithms are like solving this puzzle, calculating the joint angles that will create the desired end effector pose.

Roboticists and engineers have developed a wide range of algorithms and techniques to tackle the complexities of inverse kinematics. These algorithms vary in sophistication and are tailored to the specific needs of different robotic systems.

Let's consider an example from the field of robotics research, where a humanoid robot is designed to mimic human movements. Inverse kinematics is essential to make the robot replicate human gestures and postures accurately.

Imagine the robot attempting to wave its hand in a friendly manner. Inverse kinematics algorithms calculate the joint angles of the robot's arm and hand to achieve this waving motion. The result is a lifelike and expressive gesture that mirrors human behavior.

One fascinating aspect of forward and inverse kinematics is redundancy. Robots with more degrees of freedom, such as extra joints, can have multiple solutions to reach the same end effector position and orientation. It's like having several ways to perform the same dance move.

Consider a snake-like robot with numerous segments. It can bend and twist its body in various ways to reach a particular target. Forward and inverse kinematics help the robot choose the most suitable joint angles among the many possible solutions.

In practical applications, redundancy can be advantageous. For instance, a robot with redundancy can adapt to changing environments or avoid obstacles by selecting joint angles that optimize its movements.

In summary, forward kinematics and inverse kinematics are the choreographers of the robotic world, shaping the intricate dance of robot arms and limbs. They enable robots to perform tasks with precision and grace, whether it's assembling products, mimicking human gestures, or executing delicate surgical procedures. So, the next time you see a robot gracefully moving its limbs, remember the mathematical elegance of forward and inverse kinematics, orchestrating the magic of robotic motion in our ever-evolving technological landscape.

Welcome to the exciting world of trajectory planning and control, where robots and autonomous vehicles learn to navigate complex environments, follow precise paths, and execute intricate movements, much like skilled performers on a grand stage. In the realm of robotics and automation, trajectory planning and control are like the conductor and musicians of an orchestra, ensuring that every motion is harmonious and flawless.

Trajectory planning is the art of designing a path for a robot or vehicle to follow, taking into account its dynamics and constraints. Think of it as mapping out the choreography for a dance routine, specifying every step, turn, and leap. In the robotic world, trajectory planning guides the motion of the robot's end effector or the vehicle's wheels, creating a sequence of waypoints that define its path.

Imagine an autonomous car driving through a bustling city. Trajectory planning algorithms determine the car's route, its speed profile, and when it should turn or merge lanes. This ensures that the car navigates safely and smoothly, avoiding obstacles and adhering to traffic rules.

One of the key considerations in trajectory planning is optimizing various criteria. Robots and vehicles often aim to minimize travel time, energy consumption, or wear and tear on their components. This is like a dancer striving to make every movement efficient and graceful, conserving energy for a flawless performance.

Consider a drone delivering packages to homes. Trajectory planning algorithms calculate the drone's flight path to reach each delivery location efficiently while avoiding obstacles like trees and buildings. Optimization ensures that the drone delivers packages quickly and conserves battery power.

Dynamic environments add complexity to trajectory planning. Imagine a warehouse where autonomous robots move among human workers. Trajectory planning algorithms must account for the constantly changing positions of both robots and humans to avoid collisions and maintain smooth operations.

One fascinating aspect of trajectory planning is its adaptability. Robots and vehicles can adjust their trajectories in real-time based on sensor feedback. Think of it as a dancer improvising and adjusting movements in response to the music and audience reactions. Real-time trajectory adjustments ensure that robots and vehicles can handle unexpected situations.

Now, let's dive into trajectory control, which is responsible for executing the planned path with precision. It's like the conductor guiding the orchestra to play each note flawlessly, ensuring that the robot or vehicle follows the planned trajectory accurately.

Consider a robot arm assembling delicate electronic components. Trajectory control algorithms ensure that the arm moves smoothly and precisely, avoiding errors in placement or assembly. This level of control is crucial for tasks requiring high precision.

Trajectory control involves managing various factors, including velocity, acceleration, and torque. Robots and vehicles must adjust their speed and force to match the planned trajectory. It's like a dancer modulating their movements to match the rhythm and intensity of the music.

In autonomous vehicles, trajectory control is essential for maintaining stability and safety. For instance, when a car encounters a slippery road, its control system adjusts the wheel torque and braking to prevent skidding and maintain control. This is akin to a dancer maintaining balance on a slippery stage.

Now, let's explore feedback control, a fundamental concept in trajectory control. Feedback control systems continuously monitor the robot's or vehicle's state and make adjustments to ensure it follows the desired trajectory. It's like a conductor listening to the orchestra and making instant adjustments to keep the music on track.

Consider a drone flying in windy conditions. Feedback control algorithms analyze sensor data, such as gyroscopes and accelerometers, to detect deviations from the planned trajectory. The control system then adjusts the drone's motors to maintain stability and stay on course.

Another critical aspect of feedback control is error correction. Robots and vehicles are not perfect, and there can be discrepancies between the planned trajectory and the actual path. Feedback control systems detect these errors and apply corrective actions. It's like a dancer noticing a misstep and quickly regaining their balance to continue the performance seamlessly.

Proportional-Integral-Derivative (PID) controllers are commonly used in feedback control. They calculate control signals based on the current error, its integral (accumulated error over time), and its derivative (rate of change). PID controllers provide a balance between stability and responsiveness, much like a conductor maintaining the tempo of a musical piece.

In autonomous vehicles, feedback control is crucial for tasks like lane-keeping and adaptive cruise control. These systems use sensor data to ensure the vehicle stays within its lane and maintains a safe following distance from other vehicles.

Robotic arms in manufacturing also rely on feedback control to achieve precise movements. Feedback sensors, such as encoders and force sensors, provide real-time data about the arm's position and interactions with the environment. The control system adjusts the arm's movements to meet the desired trajectory and maintain quality in manufacturing processes.

One exciting development in trajectory planning and control is the integration of machine learning. Machine learning algorithms can adapt and improve trajectory planning and control based on experience. They can learn from past successes and failures to make better decisions.

Imagine a robot navigating a cluttered environment filled with unpredictable obstacles. Machine learning algorithms can analyze sensor data to learn how to navigate efficiently, avoid collisions, and even discover new paths through trial and error. This adaptability is like a dancer learning new moves and improvising to create unique performances.

In summary, trajectory planning and control are the choreographers and conductors of the robotic world, shaping the intricate dance of robots and autonomous vehicles as they navigate, perform tasks, and interact with their surroundings. Whether it's an autonomous car gracefully weaving through traffic or a robot arm delicately assembling intricate components, these technologies ensure precision, adaptability, and safety in the ever-evolving landscape of robotics and automation. So, the next time you witness a robot executing flawless movements or an autonomous vehicle navigating with precision, remember the artistry of trajectory planning and control, orchestrating the magic of robotic motion in our technologically driven world.

Chapter 7: Programming and Software Development for Robotics

Welcome to the realm of software architecture for robotics, where the digital brains of robots are designed and built, orchestrating their every move and decision. It's akin to the architect crafting the blueprints for a grand building, where every element of the design serves a purpose and functions seamlessly together.

In the world of robotics, software architecture serves as the foundation upon which the entire robot's functionality rests. It's the intricate web of code, algorithms, and systems that enables robots to perceive their surroundings, make decisions, and execute tasks with precision.

Imagine a robot navigating a cluttered environment, such as a warehouse filled with shelves and products. Software architecture is the guiding force that allows the robot to process data from its sensors, create a map of the environment, plan a path, and control its movements—all in real-time.

One of the fundamental principles in software architecture for robotics is modularity. This concept is similar to building blocks in construction, where each component of the software is designed as a separate module with specific functions and interfaces. These modules can be developed, tested, and maintained independently, making the software more robust and adaptable.

Consider a robotic arm in manufacturing. Its software architecture consists of modules for controlling each joint, interpreting user commands, and ensuring safety. This modular approach allows engineers to update or replace individual components without affecting the entire system.

Another critical aspect of software architecture is abstraction. Abstraction is like the artist's brushstroke, simplifying complex details and presenting a high-level view of the system. In robotics, abstraction hides the intricacies of hardware and low-level operations, allowing developers to focus on high-level tasks and algorithms.

Think of a self-driving car's software architecture. Abstraction layers hide the complexities of sensors, actuators, and hardware interfaces, providing a clean and intuitive interface for developers to work with. This abstraction enables the development of autonomous driving algorithms without diving into the nitty-gritty details of sensor calibration or motor control.

Real-time processing is a paramount requirement in many robotic applications. It's like a conductor ensuring that the orchestra plays in perfect sync with the music. In robotics, real-time software architecture guarantees that sensor data is processed and acted upon within tight time constraints.

Consider a drone flying through a forest to capture images for environmental monitoring. Real-time software architecture ensures that the drone's control algorithms react instantaneously to obstacles detected by its sensors. This responsiveness is crucial for avoiding collisions and ensuring safe flight.

In robotics, middleware acts as the communication hub that allows different software components to exchange data and commands seamlessly. Think of it as the postal service that delivers messages between different parts of the robot's brain. Middleware simplifies the development of complex robotic systems by providing standardized communication protocols.

Imagine a robot in a smart home environment. Middleware enables the robot to communicate with various devices like thermostats, lights, and security cameras. It can receive commands to adjust the room temperature, turn off the lights, or provide a live video feed—all through a unified communication framework.

Machine learning and artificial intelligence (AI) are integral parts of modern robotics software architecture. Machine learning algorithms, like neural networks, enable robots to learn from data and improve their performance over time. It's akin to the robot evolving and adapting based on its experiences.

Consider a robot in a hospital that learns to navigate busy hallways and avoid collisions with people. Machine learning algorithms analyze sensor data to identify patterns and make decisions, allowing the robot to become more efficient and safer with each interaction.

Security is a paramount concern in robotics software architecture, much like safeguarding a precious treasure. Robots often interact with sensitive data and physical environments, making them vulnerable to cyberattacks or unauthorized access. Security measures, including encryption, authentication, and access control, are crucial to protect both the robot and its surroundings.

Think of a robot in a smart manufacturing facility. Security measures ensure that only authorized personnel can access and control the robot's functions. This safeguards the integrity of the manufacturing process and protects against potential security breaches.

Another key consideration in software architecture for robotics is scalability. Scalability is like designing a building with the flexibility to accommodate future growth. In robotics, software should be able to adapt to changing requirements and hardware configurations.

Consider a fleet of autonomous delivery robots for a logistics company. Scalable software architecture allows the company to add more robots to its fleet without a complete overhaul of the software. This scalability ensures efficiency and cost-effectiveness as the business expands.

Open-source software plays a significant role in robotics software architecture. Open-source projects are like collaborative art, where developers from around the world contribute their expertise to create software that is accessible to all. Open-source robotics software enables innovation and accelerates the development of robotic applications.

Imagine a research lab working on a humanoid robot. They leverage open-source software libraries for tasks like perception, motion planning, and control. This collaborative approach saves time and resources, allowing researchers to focus on advancing the state of the art in robotics.

Interoperability is a critical aspect of software architecture for robotics. Interoperability ensures that robots can work together and exchange information seamlessly. It's like different musicians from various orchestras coming together to perform a symphony.

Consider a scenario where multiple robots collaborate to perform a complex task, such as search and rescue in a disaster-stricken area. Interoperability standards enable these robots to share sensor data, coordinate their movements, and make joint decisions, enhancing their effectiveness in critical situations.

In summary, software architecture for robotics is the unseen conductor guiding the symphony of robotic movements and decision-making. It's the framework that enables robots to perceive, think, and act in the real world. Whether it's a drone soaring through the sky, a robot assisting in surgery, or an autonomous vehicle navigating city streets, software architecture is the mastermind behind their capabilities. So, the next time you witness a robot gracefully performing a task, remember the intricate dance of software architecture that makes it all possible, orchestrating the magic of robotics in our ever-evolving technological landscape.

Welcome to the fascinating world of testing and debugging in robotics, where engineers and developers embark on a journey of discovery, troubleshooting, and refinement to ensure that robots perform flawlessly in the real world. Think of it as the quest to make sure that our robotic creations are as dependable as a trusted friend.

In the realm of robotics, testing is the process of subjecting robots to a series of carefully designed experiments and scenarios to evaluate their performance, identify issues, and gather data for improvement. It's like taking a new car for a test drive to ensure it handles various road conditions without a hitch.

Consider a robot designed to navigate an unfamiliar environment. Testing involves placing the robot in different situations, such as crowded spaces, uneven terrain, or low-light conditions, to assess its ability to adapt and make informed decisions. The goal is to uncover any weaknesses and improve the robot's performance.

Testing in robotics often starts with simulation. Simulation environments allow developers to create virtual worlds where robots can be tested without physical hardware. It's like a digital playground where robots can learn and refine their skills without any risk of damage.

Imagine a team of engineers developing an autonomous drone. They use a simulation environment to mimic various flight scenarios, such as windy conditions or obstacle-rich environments. By conducting thousands of simulated flights, they can fine-tune the drone's algorithms and ensure it can handle diverse challenges.

Once a robot's software and algorithms have undergone simulation testing, it's time for real-world testing. This is where the robot steps out of the virtual realm and into the physical world. Real-world testing is akin to a dress rehearsal for a live performance, where the robot faces the unpredictability and complexity of the real environment.

Consider a self-driving car undergoing real-world testing on city streets. Engineers equip the car with sensors and software, and it navigates through traffic, interacts with pedestrians, and adheres to traffic laws. This phase of testing provides valuable insights into how the car performs in dynamic and uncontrolled settings.

Testing is not a one-time event but an iterative process. Engineers gather data from each test, analyze the results, and make adjustments to improve the robot's performance. It's like a painter refining a masterpiece with each brushstroke, striving for perfection.

Now, let's delve into the art of debugging in robotics. Debugging is the detective work of identifying and rectifying issues or bugs in a robot's software or hardware. It's akin to solving a puzzle where each bug represents a missing piece that needs to be found and placed in its proper spot.

Imagine a robotic arm in a manufacturing facility that occasionally fails to grasp objects securely. Debugging involves analyzing sensor data, reviewing code, and conducting experiments to pinpoint the source of the issue. It's a process of trial and error to ensure the robot's reliability.

One essential tool in debugging is logging. Logging involves recording detailed information about the robot's behavior, sensor readings, and software execution. It's like keeping a journal of the robot's actions, providing a trail of clues for developers to follow when issues arise.

Consider a humanoid robot that sometimes loses balance while walking. Logging captures data on the robot's joint angles, sensor feedback, and foot placements. By analyzing this data, engineers can identify patterns and anomalies that help them diagnose and resolve the balance problem.

Debugging also involves the use of debugging tools and software. These tools allow developers to step through code, set breakpoints, and inspect variables in real-time. It's like having a magnifying glass to examine the details of a complex puzzle.

Imagine a team of programmers debugging the control software for a swarm of small robots. Debugging tools enable them to pause the simulation, examine the robots' states, and modify code on-the-fly to address issues related to coordination and communication.

One of the challenges in debugging robotics systems is dealing with intermittent issues that occur sporadically and are challenging to reproduce consistently. It's like trying to catch a mischievous ghost that only appears when you least expect it.

Consider a drone that occasionally experiences GPS signal loss, causing it to veer off course. Debugging such intermittent issues requires patience and perseverance, as developers must analyze data from multiple flight sessions to identify patterns and root causes.

Collaboration is a key aspect of effective debugging in robotics. Debugging often involves interdisciplinary teams, including mechanical engineers, software developers, and hardware specialists. It's like assembling a group of experts to solve a complex puzzle together.

Imagine a robot designed for underwater exploration encountering problems with its sensors. Collaboration between the robot's mechanical engineers, who assess the physical components, and its software developers, who examine the algorithms, is essential to diagnose and resolve the issues.

Regression testing is another critical practice in robotics debugging. When changes are made to a robot's software or hardware, regression testing ensures that previously resolved issues do not reappear. It's like ensuring that a fixed leaky roof remains watertight after a storm.

Consider a robotic vacuum cleaner that receives a software update to improve its navigation. Regression testing involves running a battery of tests to verify that the update does not reintroduce problems related to getting stuck or missing spots.

One fascinating aspect of debugging in robotics is the use of advanced technologies like remote debugging and teleoperation. These technologies allow developers to debug robots from a distance, even when they are deployed in remote or hazardous environments.

Imagine a rover exploring the surface of Mars. Remote debugging enables engineers on Earth to connect to the rover, examine its software and sensors, and diagnose issues without the need for physical access. This capability is crucial for maintaining the rover's functionality in the harsh Martian terrain.

In summary, testing and debugging are the twin pillars of ensuring that robots are not just technological marvels but reliable and trustworthy companions in various domains, from manufacturing floors to outer space exploration. Testing ensures that robots can perform their tasks under diverse conditions, while debugging uncovers and fixes issues to make them more dependable. So, the next time you witness a robot flawlessly executing a task, remember the behind-the-scenes work of testing and debugging that made it all possible, ensuring that our robotic companions are always ready for their performance in the world's grand stage.

Chapter 8: Machine Learning and AI in Robotics

Welcome to the exciting world of machine learning in robotics, where algorithms learn from data and experience, enabling robots to adapt, make decisions, and interact with their environments just like a seasoned explorer navigating uncharted territory. Think of it as the art of instilling intelligence into machines, allowing them to perceive and respond to the world around them.

In the realm of robotics, machine learning serves as the brainpower that empowers robots to tackle complex tasks with precision and efficiency. It's like giving a robot the ability to learn, reason, and improve its performance, much like we humans do as we gain experience.

Consider a robot designed to assist in a hospital by delivering medication to patients. Machine learning algorithms can be employed to allow the robot to navigate the hospital's dynamic environment, avoiding obstacles, and ensuring timely and safe deliveries.

At its core, machine learning involves the creation of algorithms that can automatically improve their performance over time through learning from data. This process is akin to nurturing a young mind, where the robot becomes more adept at its tasks as it gathers experience.

Think of a self-driving car that learns from millions of miles of driving data. Machine learning algorithms analyze this data to improve the car's ability to recognize objects, predict traffic patterns, and make split-second decisions, ultimately making autonomous driving safer and more reliable.

Machine learning in robotics relies heavily on the concept of pattern recognition. Robots are trained to recognize patterns in data, such as images, sensor readings, or even spoken language. It's like teaching a robot to understand the language of its surroundings.

Imagine a robot tasked with sorting packages in a logistics center. Machine learning enables the robot to recognize the unique patterns on each package label, allowing it to sort packages accurately and efficiently.

One of the fundamental techniques in machine learning is supervised learning. In supervised learning, robots are provided with labeled training data, consisting of input-output pairs. It's like a teacher guiding a student, providing examples and corrections.

Consider a robot designed to pick and place objects in a warehouse. Supervised learning allows the robot to learn how to grasp different objects by observing human demonstrations and receiving feedback on its actions.

Unsupervised learning is another key technique in machine learning for robotics. In unsupervised learning, robots analyze data without explicit labels, seeking to discover hidden patterns and structures. It's like a robot exploring a new environment, making sense of what it encounters.

Imagine a robot tasked with exploring the ocean floor to study marine life. Unsupervised learning algorithms help the robot identify clusters of similar organisms or geological features, contributing to scientific discoveries.

Reinforcement learning adds a dynamic element to the machine learning landscape. In reinforcement learning, robots learn by interacting with their environment and receiving rewards or penalties based on their actions. It's like a robot playing a game, where it strives to maximize its score over time.

Consider a drone learning to navigate through a forest to locate lost hikers. Reinforcement learning enables the drone to adapt its flight path based on the terrain it encounters and the feedback it receives, ultimately improving its search and rescue capabilities.

Machine learning in robotics is not limited to just algorithms. It often involves the integration of various sensors, such as cameras, lidar, and inertial measurement units, to gather data about the robot's surroundings. It's like providing the robot with eyes, ears, and a sense of touch.

Think of a humanoid robot in a human-robot collaboration scenario. Machine learning algorithms process data from the robot's cameras and force sensors, allowing it to detect human movements and adjust its actions accordingly, ensuring safe and effective collaboration.

One of the remarkable aspects of machine learning is its ability to adapt and generalize. Robots can learn from one task or environment and apply that knowledge to new, unseen situations. It's like a robot gaining expertise in one field and then effortlessly transferring its skills to another.

Imagine a robot initially trained to recognize and sort fruit in a grocery store. With transfer learning, the same robot can be deployed in a warehouse to sort a different set of items, such as electronics, by leveraging the knowledge it gained from its previous task.

In robotics, machine learning also plays a crucial role in perception and understanding the world. Computer vision, a subfield of machine learning, equips robots with the ability to interpret and make sense of visual data. It's like giving a robot the gift of sight.

Consider a robot in a manufacturing facility that uses computer vision to inspect products for defects. Machine learning algorithms analyze images of the products, allowing the robot to identify even subtle flaws and take appropriate actions, such as flagging or removing defective items.

Natural language processing (NLP) is another dimension of machine learning in robotics. It enables robots to understand and generate human language, facilitating communication and interaction. It's like giving a robot the power to converse and comprehend our words.

Imagine a robot in a customer service role, assisting customers by answering questions and providing information. NLP algorithms empower the robot to understand spoken or typed queries and respond in a natural and human-like manner, enhancing the user experience.

Machine learning in robotics is not without its challenges. Robots must learn efficiently, adapt to changing environments, and operate safely. Moreover, they must be able to handle uncertainty and make decisions in real-time.

Consider a robot exploring a disaster-stricken area to locate survivors. Machine learning algorithms must enable the robot to navigate debris, make rapid decisions based on incomplete information, and avoid dangerous situations, all while prioritizing human safety.

In summary, machine learning is the secret sauce that transforms robots from mere machines into intelligent, adaptable companions in our modern world. It's the engine that drives their ability to learn, reason, and interact with us and our environments. So, the next time you see a robot effortlessly navigating a complex task or understanding your commands, remember the magic of machine learning that powers its capabilities, making it an invaluable member of our technological landscape.

Welcome to the captivating realm of deep learning for robot perception, a domain where machines learn to perceive and understand the world around them with remarkable depth and precision. Think of it as the art of bestowing robots with a level of perception that approaches our own, enabling them to interpret visual and sensory data with astonishing accuracy.

In the vast landscape of robotics, perception is the foundation upon which intelligent decision-making is built. It's akin to providing robots with the ability to see, hear, and feel their surroundings, allowing them to navigate complex environments and interact with objects and people.

Consider a robot designed for warehouse logistics. Deep learning equips it with the ability to identify products on shelves, recognize labels, and determine the best way to grasp and transport items, just as a human worker would.

Deep learning, a subset of machine learning, delves into the intricacies of neural networks inspired by the human brain. These artificial neural networks are composed of layers of interconnected nodes, known as neurons, that process data hierarchically. It's like building a complex puzzle where each piece (neuron) contributes to the overall picture.

Imagine an image recognition system within a robot's camera. Deep learning neural networks dissect the image into layers of features, starting with edges and textures and progressing to complex objects and scenes. This hierarchical approach enables robots to perceive the world in a manner that mimics human vision.

Convolutional Neural Networks (CNNs) are a prominent architecture within deep learning that excels in tasks involving images and visual data. They operate by applying convolutional filters to input data to extract meaningful features. It's like a robot's way of dissecting an image into its constituent elements.

Consider a self-driving car's perception system. CNNs analyze the car's camera feed to detect lane markings, traffic signs, pedestrians, and other vehicles. This enables the car to make real-time decisions about lane changing, stopping, or yielding, ensuring safe navigation.

Recurrent Neural Networks (RNNs) are another facet of deep learning that specializes in sequential data, making them suitable for tasks involving time-series data or sequences of events. It's like giving a robot a memory of past experiences.

Imagine a robot assisting in healthcare by monitoring a patient's vital signs over time. RNNs enable the robot to analyze a continuous stream of data, such as heart rate and blood pressure readings, to detect trends or anomalies that may require medical attention.

Generative Adversarial Networks (GANs) introduce a fascinating dynamic to deep learning. GANs consist of two neural networks, a generator and a discriminator, engaged in a competitive dance. The generator attempts to create realistic data, while the discriminator tries to distinguish between real and generated data. It's like a robot honing its ability to generate lifelike images or sounds.

Consider a robot artist creating original pieces of art. GANs allow the robot to generate unique artworks by learning from a dataset of artistic styles and motifs. The generator produces art, while the discriminator evaluates its authenticity and provides feedback for improvement.

Transfer learning is a valuable concept within deep learning, enabling robots to leverage knowledge gained from one domain to excel in another. It's like a versatile athlete applying skills from one sport to excel in another.

Imagine a robot initially trained for object recognition in images. Through transfer learning, the same robot can adapt its knowledge to perform tasks like detecting defects in manufactured products, leveraging its understanding of visual patterns.

One of the remarkable facets of deep learning for robot perception is its ability to handle unstructured data. This includes not only images but also audio, text, and even sensor data. It's like equipping robots with multiple senses.

Consider a robot designed for natural language understanding. Deep learning models process spoken or written text, enabling the robot to comprehend and respond to human commands or engage in conversations, making it an effective virtual assistant.

Multi-modal perception is an exciting frontier in deep learning for robotics. It involves integrating data from multiple sources, such as cameras, microphones, and touch sensors, to build a comprehensive understanding of the environment. It's like a robot becoming more perceptive by combining its various senses.

Imagine a robot designed to assist people with visual impairments. Multi-modal perception allows the robot to not only detect obstacles through sensors but also provide audio descriptions of the environment, enhancing the user's mobility and safety.

Anomaly detection is a critical application of deep learning in robot perception. Robots equipped with deep learning models can identify deviations from normal patterns in data, helping detect faults or unusual events. It's like a robot acting as a vigilant guardian, watching for signs of trouble.

Consider a manufacturing robot tasked with assembling electronic components. Anomaly detection algorithms monitor the quality of each component's placement and soldering, flagging any deviations from the expected standards to prevent defects.

Continual learning is an emerging area in deep learning for robotics. It focuses on enabling robots to adapt to changing environments and evolving tasks over time. It's like a robot that not only learns but also grows and evolves with experience.

Imagine a delivery robot operating in a dynamic urban environment. Continual learning allows the robot to adapt to new traffic patterns, road closures, and pedestrian behavior, ensuring efficient and safe deliveries.

Despite the incredible potential of deep learning for robot perception, challenges remain. Deep learning models require significant computational resources and large datasets for training. Moreover, ensuring the safety and reliability of robots equipped with perception systems is an ongoing concern.

Consider a robot in a healthcare setting responsible for dispensing medication to patients. Ensuring that the robot's perception system can reliably identify patients and administer the correct medication is a matter of utmost importance.

In summary, deep learning for robot perception is a frontier where machines strive to perceive and understand the world with increasing sophistication. It's the journey of imbuing robots with the ability to see, hear, and interpret their surroundings, allowing them to excel in a wide range of applications, from autonomous vehicles to healthcare assistants. So, the next time you witness a robot effortlessly identifying objects or recognizing speech, remember the incredible depth of perception made possible by the magic of deep learning, shaping the future of robotics and technology.

Chapter 9: Sensor Fusion and Multi-Sensor Integration

Welcome to the fascinating world of sensor data fusion techniques, where the art of blending information from various sensors creates a comprehensive and accurate picture of the world around us. Think of it as orchestrating a symphony of sensory inputs, enabling robots and autonomous systems to make informed decisions and navigate complex environments.

In the realm of robotics and autonomous vehicles, sensors serve as the eyes, ears, and touch of machines. They capture critical information about the environment, such as spatial data, temperature, and even chemical compositions, allowing machines to interact with their surroundings effectively.

Imagine an autonomous car navigating through a bustling city. Sensor data fusion techniques ensure that the car's multiple sensors, including cameras, lidar, radar, and ultrasonic sensors, work together seamlessly to provide a holistic view of the road, other vehicles, pedestrians, and obstacles.

The primary goal of sensor data fusion is to enhance the accuracy and reliability of the information obtained from sensors. It's like having a team of experts collaborate to provide the most precise insights.

Consider a military drone conducting reconnaissance missions. Sensor data fusion techniques enable the drone to combine information from various sensors, such as thermal imaging cameras and radio frequency detectors, to detect and identify potential threats with high confidence.

One of the fundamental approaches to sensor data fusion is known as sensor fusion or multi-sensor fusion. It involves integrating data from multiple sensors to create a more informative and robust representation of the environment. It's like combining puzzle pieces to reveal the complete picture.

Imagine a search and rescue robot deployed in a disaster-stricken area. Sensor fusion allows the robot to merge data from its visual cameras, infrared sensors, and acoustic detectors to locate survivors trapped under debris accurately.

Sensor data fusion techniques can be categorized into several types, each suited to specific applications and scenarios. One common approach is sensor-level fusion, where raw data from sensors are combined directly. It's like assembling ingredients to create a delicious dish.

Consider a weather monitoring station that combines data from various sensors, such as anemometers, barometers, and thermometers, to provide accurate and up-to-date weather forecasts for a region.

Feature-level fusion takes a more advanced approach by extracting relevant features or patterns from sensor data before fusion. It's like distilling the essence of a complex flavor to create a unique blend.

Imagine an agricultural robot tasked with identifying and harvesting ripe fruit. Feature-level fusion enables the robot to extract color and texture features from its cameras to distinguish between ripe and unripe fruit accurately.

Decision-level fusion involves combining the decisions or outputs of individual sensors to make a final decision or estimation. It's like having multiple experts provide their opinions, and a committee makes the ultimate call.

Consider a security system in a smart home that combines decisions from motion detectors, door sensors, and surveillance cameras to determine whether to alert the homeowner of a potential intrusion.

Sensor data fusion techniques are not limited to just combining data from the same type of sensors. They can also fuse data from heterogeneous sensors, such as combining visual and acoustic data. It's like understanding a story from both the written text and an audiobook.

Imagine a wildlife monitoring system that fuses data from cameras and acoustic sensors to detect and track elusive animals based on both visual and acoustic cues.

One of the exciting aspects of sensor data fusion is the ability to handle uncertainty. Sensors may provide imperfect or noisy data, and fusion techniques help mitigate these uncertainties. It's like having a team of detectives cross-checking their findings to arrive at the most accurate conclusion.

Consider a robot exploring the seafloor for archaeological research. Sensor data fusion techniques allow the robot to filter out noise and anomalies in its sonar and camera data, ensuring precise mapping and artifact detection.

Sensor data fusion also plays a crucial role in localization and mapping, known as simultaneous localization and mapping (SLAM). Robots use sensor data to navigate and build maps of their surroundings in real time. It's like a cartographer creating a detailed map as they explore uncharted territories.

Imagine an autonomous drone mapping a dense forest. SLAM techniques enable the drone to combine data from its GPS, lidar, and onboard cameras to create a 3D map of the forest, enabling accurate navigation and obstacle avoidance.

One of the key challenges in sensor data fusion is handling conflicting information from sensors. In some cases, sensors may provide contradictory data, and fusion techniques must resolve these conflicts intelligently. It's like having referees in a sports game who must make fair decisions even when faced with conflicting reports from players.

Consider a self-driving car receiving data from both its lidar and radar sensors. Sensor data fusion techniques must reconcile any discrepancies in the measurements to make safe driving decisions.

Sensor data fusion extends beyond robotics and autonomous systems. It finds applications in fields such as healthcare, where it can integrate data from various medical sensors to provide comprehensive patient monitoring and diagnosis.

Imagine a wearable health device that fuses data from sensors measuring heart rate, body temperature, and blood oxygen levels to provide real-time health insights and early warning of potential health issues.

Another exciting application is in environmental monitoring, where sensor data fusion can combine data from sensors measuring air quality, temperature, humidity, and pollutant levels to provide a holistic view of environmental conditions.

Consider an urban planning project that uses sensor data fusion to monitor air quality in a city. The integrated data can help policymakers make informed decisions to improve air quality and public health.

In summary, sensor data fusion techniques are the orchestrators of harmony in the world of sensors and autonomous systems. They enable machines to make sense of the vast array of sensory inputs, providing a comprehensive and accurate understanding of the environment. So, the next time you witness a self-driving car navigating effortlessly through traffic or a search and rescue robot finding a trapped survivor, remember the intricate dance of sensor data fusion that makes it all possible, shaping the future of technology and our interaction with the world.

Welcome to the captivating world of localization and mapping with multiple sensors, a realm where machines not only navigate with precision but also construct detailed maps of their surroundings in real-time. Think of it as the art of digital exploration, where robots and autonomous systems venture into unknown territories, finding their way and creating maps as they go.

Localization, in the context of robotics, is all about determining a machine's position in a known environment. It's akin to a treasure hunt, where the machine seeks to pinpoint its location on a digital map with utmost accuracy. This capability is essential for various applications, from self-driving cars navigating city streets to drones exploring remote landscapes.

Imagine a delivery robot tasked with delivering packages to specific locations in a busy urban area. Localization techniques ensure that the robot knows precisely where it is at all times, allowing it to reach its destinations efficiently.

Global Positioning System (GPS) is one of the most well-known localization technologies. It uses signals from satellites to determine a device's location on Earth. However, GPS has limitations, especially in urban canyons or indoor environments, where signals may be obstructed or inaccurate.

Consider a construction site where heavy machinery operates amidst tall buildings. GPS alone may not provide the required level of accuracy for precise equipment positioning, making additional localization methods essential.

This is where multiple sensors come into play. Robots and autonomous systems often rely on a combination of sensors to enhance their localization capabilities. It's like having a team of scouts who use various methods to pinpoint their location in the wilderness.

Imagine a robot exploring the surface of Mars. In addition to GPS, it uses onboard cameras, accelerometers, and gyroscopes to navigate the Martian terrain accurately and avoid obstacles.

One of the exciting challenges in localization is fusing data from multiple sensors to achieve high accuracy. Sensor fusion, a concept we explored earlier, combines data from sensors like cameras, lidar, radar, and inertial sensors to create a more comprehensive understanding of the environment.

Consider a self-driving car on a highway. It fuses data from its GPS, lidar, and camera sensors to not only determine its precise location but also detect other vehicles and obstacles on the road.

Mapping, on the other hand, is the art of creating a digital representation of the environment. It's like drawing a detailed map of an intricate labyrinth as you explore its twists and turns.

Imagine an underwater remotely operated vehicle (ROV) exploring the depths of the ocean. As it moves through underwater caves and trenches, it simultaneously constructs a 3D map of the seabed, enabling scientists to study marine ecosystems.

One of the groundbreaking developments in mapping technology is Simultaneous Localization and Mapping, or SLAM. SLAM is like a robotic cartographer that not only determines the robot's location but also builds a map of the environment on-the-fly.

Imagine a warehouse robot tasked with efficiently transporting goods. As it moves through the facility, SLAM algorithms enable it to create a map of the warehouse layout and accurately position itself for tasks like picking and delivering items.

One essential aspect of SLAM is loop closure, which ensures that the robot revisits previously mapped areas and closes the loop, eliminating errors in its estimated position. It's like retracing your steps to complete a complex maze accurately.

Consider an autonomous vacuum cleaner in a multi-story building. Loop closure ensures that the vacuum cleaner doesn't miss any areas while navigating between floors, resulting in thorough cleaning.

Multiple sensors play a vital role in SLAM. Robots often employ a combination of laser-based lidar sensors, visual cameras, and inertial measurement units (IMUs) to gather data about their surroundings. It's like having a team of specialists, each contributing their unique expertise to solve a complex puzzle.

Imagine a drone mapping a forested area for conservation research. The drone uses lidar to measure tree heights, cameras to capture high-resolution images, and IMUs to track its flight path, enabling the creation of detailed forest maps.

One of the remarkable applications of SLAM is in the field of autonomous exploration. Robots equipped with SLAM capabilities can venture into unknown or hazardous environments, mapping as they go, and providing valuable data without human intervention.

Consider a planetary rover exploring the surface of Mars. Using SLAM, it can autonomously navigate the Martian terrain, avoiding obstacles, and creating maps that help scientists plan future missions.

Another fascinating aspect of mapping with multiple sensors is multi-modal mapping. This approach involves combining data from different types of sensors, such as visual and lidar data, to create more informative maps. It's like building a puzzle with pieces of varying shapes and colors.

Imagine an agricultural robot monitoring crop health. Multi-modal mapping allows the robot to fuse data from cameras and hyperspectral sensors, enabling it to create maps that not only show the field's layout but also highlight areas with specific crop conditions.

The fusion of localization and mapping with multiple sensors extends beyond robotics into various domains. In augmented reality applications, such as mobile gaming, smartphones use a combination of GPS, accelerometers, and cameras to enable users to explore virtual worlds superimposed on the real environment.

Consider a mobile game where players hunt for virtual creatures in their city. The game uses GPS for location-based gameplay, accelerometers for tracking movement, and cameras for capturing augmented reality creatures in the real world.

In scientific research, remote sensing satellites equipped with various sensors capture data about Earth's surface and atmosphere. By fusing data from sensors like optical cameras and microwave radiometers, scientists gain insights into climate change, land use, and natural disasters.

Imagine a satellite orbiting the Earth and collecting data about the planet's temperature, humidity, and vegetation cover. Multi-sensor fusion allows scientists to create comprehensive maps and models for environmental studies.

In the world of autonomous vehicles, localization and mapping with multiple sensors are pivotal for achieving safe and efficient transportation. Self-driving cars, for instance, rely on a combination of GPS, lidar, radar, and cameras to navigate city streets and highways.

Consider a future where self-driving cars seamlessly navigate through complex urban environments. Multiple sensors working in harmony provide these vehicles with the ability to accurately determine their position, detect other vehicles, pedestrians, and road conditions, ensuring safe and reliable transportation.

In summary, the synergy of localization and mapping with multiple sensors opens the door to a world where machines can explore, navigate, and understand their surroundings with remarkable accuracy. Whether it's a delivery robot finding its way through a bustling city or a planetary rover mapping the rugged terrain of another world, the fusion of sensor data propels us into a future where machines and technology continuously push the boundaries of exploration and discovery. So, the next time you witness a robot effortlessly creating maps or a self-driving car navigating a complex intersection, marvel at the intricate dance of sensors and data fusion that makes it all possible, shaping the way we interact with our world.

Chapter 10: Building and Testing Your First Robot

Navigating the process of selecting hardware components for your robot can be both exhilarating and challenging, akin to choosing the right tools for a creative project. Each component plays a crucial role in defining your robot's capabilities and performance, and making informed choices is essential to bring your robotic vision to life.

Picture yourself at a workbench, surrounded by a myriad of sensors, actuators, and computing devices, each offering unique functionalities. The first step in this hardware adventure is to define your robot's purpose and tasks. What will your robot do? Will it explore unknown terrains, assist with household chores, or perform precise manufacturing tasks?

Once you have a clear understanding of your robot's mission, you can start assembling the hardware components that align with your goals. Let's embark on this journey by exploring some key considerations and components to keep in mind.

The brain of your robot is undoubtedly one of the most critical components. It's like the control center that processes information and makes decisions. You'll need a microcontroller or a single-board computer to serve as the brains behind your robot.

Microcontrollers, such as Arduino and Raspberry Pi, are popular choices for smaller and less complex robots. They are cost-effective and energy-efficient, making them suitable for tasks like sensor data processing and basic control.

Imagine building a small line-following robot that navigates a predefined path. A microcontroller like Arduino can handle the sensor inputs and motor controls required for such a task.

For more advanced robots with complex algorithms, computer vision, or machine learning capabilities, a single-board computer like the Raspberry Pi or NVIDIA Jetson might be the better choice. These devices offer more processing power and memory, enabling your robot to handle intricate tasks.

Consider a robot designed for image recognition and object manipulation. A single-board computer can run sophisticated algorithms for object detection and manipulation, enhancing its capabilities.

Now, let's talk about sensors—the eyes and ears of your robot. Sensors are essential for gathering data about the robot's environment. The type and number of sensors you choose depend on your robot's intended tasks.

Ultrasonic sensors, for instance, are like the bat's echolocation, allowing your robot to measure distances to objects and avoid collisions. They are commonly used in robotics for obstacle detection and avoidance.

Imagine a robot designed to navigate cluttered environments, like a vacuum cleaner robot. Ultrasonic sensors help it detect obstacles and adjust its path accordingly.

Cameras are another valuable sensory tool for robots, providing visual perception. You can choose from a variety of cameras, including standard webcams and specialized cameras like the Raspberry Pi Camera Module.

For a robot that needs to recognize objects, identify faces, or perform image-based tasks, a camera is crucial. Think about a robot designed to assist in search and rescue operations. Equipped with a camera, it can locate and identify survivors in disaster-stricken areas.

Lidar sensors, often used in autonomous vehicles, emit laser beams to measure distances and create detailed 2D or 3D maps of the surroundings. They are ideal for applications requiring precise mapping and navigation.

Imagine a self-driving car navigating through a complex urban environment. Lidar sensors provide the car with accurate data about its surroundings, enabling it to make real-time decisions.

Beyond sensors, let's discuss actuators—the muscles of your robot. Actuators are responsible for making your robot move and interact with its environment. The most common types of actuators are motors.

DC motors are simple and versatile, suitable for a wide range of robotic applications. They provide rotational motion and are often used in wheeled robots for driving and steering.

Consider a robot designed for warehouse logistics, moving goods from one place to another. DC motors in the robot's wheels allow it to navigate through the facility and transport items efficiently.

Servo motors are like precision instruments, offering controlled and precise rotational movement. They are commonly used in robotic arms and manipulators for tasks requiring accuracy and fine control.

Imagine a robotic arm used in manufacturing to assemble intricate electronic components. Servo motors enable the precise positioning and movement required for delicate tasks.

Linear actuators provide linear motion, making them suitable for applications where linear movement is essential. They are often used in robotics for tasks like lifting and pushing objects.

Think about a robot designed for healthcare, assisting patients with mobility. Linear actuators can be used to lift patients in and out of bed, providing valuable support to healthcare professionals.

Wheeled robots are excellent for navigation on smooth surfaces, while tracked robots offer better traction and stability on rough terrain. Consider the environment your robot will operate in and choose the appropriate mobility system.

For example, a warehouse robot that moves packages within a controlled indoor environment may benefit from wheels for smooth and efficient navigation. On the other hand, a search and rescue robot designed to traverse uneven and challenging terrain might require tracks for stability and maneuverability.

Legged robots, inspired by nature, offer unique advantages such as agility and the ability to navigate complex environments. These robots mimic the locomotion of animals like insects or quadrupeds, making them suitable for applications where wheeled or tracked robots face limitations.

Imagine a robot designed for exploring disaster-stricken areas. A legged robot can navigate through debris and rough terrain, providing crucial assistance in search and rescue operations.

Whegs, a combination of wheels and legs, offer versatility in movement. These robots can roll like a wheel for speed or use their legs for maneuvering through tight spaces or rough terrain.

Consider a robot designed for exploring alien landscapes on other planets. Whegs provide adaptability, allowing the robot to move efficiently across various terrains.

Selecting the right power source is crucial for ensuring your robot can operate effectively. Batteries are the most common choice for mobile robots, providing portable and rechargeable energy.

The choice of battery depends on factors like your robot's power requirements and desired runtime. For smaller robots, lithium-ion or lithium-polymer batteries are often used due to their energy density and lightweight properties.

Imagine a small surveillance robot that needs to operate quietly and unobtrusively for extended periods. Lithium-polymer batteries provide the required energy density and long runtime.

For larger robots with higher power demands, such as industrial or agricultural robots, lead-acid batteries or custom battery packs may be necessary to provide the required energy. Consider an agricultural robot used for automated crop harvesting. Custom battery packs can supply the high power needed for heavy-duty tasks like harvesting and transportation.

Safety is paramount when working with robots, and this includes electrical safety. Ensure that your robot's electrical components, such as wiring and connectors, adhere to safety standards to prevent electrical hazards.

Proper insulation and shielding are essential to protect against electrical faults and interference. Regularly inspect and maintain the electrical components of your robot to ensure safe operation.

Imagine a robot used in a laboratory setting, handling hazardous materials. Electrical safety measures are critical to protect both the robot and the environment from potential risks.

In summary, selecting hardware components for your robot is a thrilling journey that allows you to shape the capabilities and personality of your mechanical creation. Whether you're designing a robot to explore distant planets, assist with everyday tasks, or tackle complex industrial challenges, the careful selection of microcontrollers, sensors, actuators, mobility systems, and power sources is key to bringing your robotic vision to fruition. So, take your time, explore the possibilities, and embrace the art of creating intelligent machines that navigate, perceive, and interact with the world around them. Your choices today will determine the abilities and achievements of the robots of tomorrow, contributing to the ever-evolving landscape of robotics and automation.

Now that we've delved into the exciting realm of selecting the essential hardware components for your robot, it's time to take the next step in your robotics journey: assembling your robot platform. Imagine this as the phase where you bring all the chosen components together, like a conductor orchestrating an ensemble, to create a harmonious and functional robotic entity.

Picture yourself in your workspace, surrounded by a collection of sensors, motors, microcontrollers, and a variety of hardware pieces. Each component represents a vital cog in the intricate machinery that is your robot, and your task is to assemble them in a way that aligns with your robot's intended purpose and design.

Before we dive into the nitty-gritty details of assembly, let's consider an essential preliminary step: planning. Just like an architect sketches the blueprint of a building before construction begins, planning is crucial for a successful robot assembly.

Start by reviewing your robot's design and functionality goals. How will the components fit together? Where will the sensors be positioned for optimal perception? How will the actuators and motors be connected to ensure efficient movement?

Planning not only helps you visualize the final product but also minimizes unexpected surprises during assembly. It's a roadmap that guides you through the intricate process of bringing your robot to life.

Assembling your robot begins with the chassis or frame—a skeleton that provides structure and support. Depending on your robot's design and mobility requirements, you may choose from various types of chassis, such as wheeled, tracked, legged, or custom-made frames.

For a wheeled robot, attach the wheels securely to the chassis using mounting brackets or hubs. Ensure that the wheels are aligned and parallel for balanced movement. If your robot has tracks, affix them carefully to the drive mechanism, ensuring proper tension for stability.

Imagine you're building a delivery robot with four wheels. Properly securing the wheels to the chassis ensures that your robot can navigate smoothly and deliver packages without any hiccups.

If your robot employs legs for locomotion, attach them in a manner that mimics the desired gait. Legged robots often require precise alignment and joint assembly to achieve stability and agility in movement.

Consider a spider-like robot designed for exploring tight spaces. The meticulous assembly of its legs allows it to traverse confined areas with ease.

Once the chassis and mobility system are in place, it's time to integrate the sensors. Carefully position the sensors to maximize their field of view and accuracy. Secure them firmly to the chassis or mounting brackets, ensuring they remain stable during robot operation.

For example, if your robot relies on a camera for vision, position it at a height and angle that provides a clear view of the surroundings. Proper sensor placement is essential for reliable data collection.

Next, connect the sensors to the microcontroller or single-board computer using appropriate cables or connectors. Ensure that the connections are secure and that you've followed the manufacturer's guidelines for sensor integration.

Imagine a robot designed for environmental monitoring, equipped with various sensors to measure temperature, humidity, and air quality. Properly connecting and integrating these sensors allows the robot to collect accurate data for analysis.

Now, let's focus on the actuators and motors. If your robot has wheels or tracks, connect the motors to the chassis or drive mechanism, ensuring they are properly aligned with the wheels or tracks. Secure them in place to prevent any unwanted movement during operation.

For a robot with a robotic arm or manipulator, assemble the joints and linkages carefully to ensure smooth and precise movement. Calibrate the actuators to achieve the desired range of motion.

Consider a robotic arm used in manufacturing to assemble products. Accurate assembly of the arm's joints and calibration of the actuators enable it to perform intricate assembly tasks with precision.

The brain of your robot, whether it's a microcontroller or single-board computer, needs a secure and accessible location on the chassis. Mount it in a way that allows for easy access to power sources, data cables, and other connections.

Ensure that the microcontroller or single-board computer is adequately cooled to prevent overheating during prolonged operation. You may need to attach heat sinks or cooling fans, depending on the computing requirements of your robot.

Imagine a robot designed for autonomous exploration in extreme environments. Proper mounting and cooling of the computer ensure that the robot can function reliably in challenging conditions.

Now, let's discuss power distribution. Your robot needs a stable and reliable power source to operate effectively. Connect the batteries or power supply unit to the microcontroller, motors, sensors, and any other power-hungry components.

Implement a power distribution system that includes voltage regulators and power management circuits to prevent voltage spikes or drops that could damage your robot's components.

Consider a robot designed for outdoor tasks, such as agricultural work. A robust power distribution system ensures that the robot can operate for extended periods without interruptions.

As you finalize the hardware assembly, take time to tidy up the wiring and cables. Organize them neatly to prevent tangling or interference with moving parts. Use cable ties or cable management accessories to secure the wires in place.

Proper cable management not only improves the aesthetics of your robot but also reduces the risk of electrical issues caused by loose or tangled wires.

Imagine a robot designed for public events, such as providing information to attendees. Neatly organized cables and wiring ensure that the robot presents a clean and professional appearance to the public.

Before powering up your robot, double-check all connections and components to ensure everything is securely in place. Perform a visual inspection to confirm that there are no loose parts or potential safety hazards.

Once you're confident in the assembly, power up your robot and test its basic functionalities. Verify that the sensors respond correctly, the motors move as expected, and the microcontroller or single-board computer operates without errors.

Consider a robot designed for educational purposes, teaching students about robotics. Thorough testing ensures that the robot functions flawlessly during demonstrations and lessons.

As your robot comes to life, document the assembly process and create a comprehensive guide. This documentation will be invaluable for troubleshooting, maintenance, and future modifications.

Imagine sharing your robot design with others in the robotics community. Your detailed documentation makes it easier for fellow enthusiasts to replicate and build upon your work.

In summary, assembling your robot platform is an essential phase in the journey of creating a functional and purposeful robotic system. By carefully planning, assembling the chassis, integrating sensors and actuators, connecting the brain, managing power distribution, and ensuring tidy wiring, you pave the way for a successful robot assembly. Each step you take brings you closer to seeing your robotic creation in action, whether it's exploring unknown terrain, assisting with tasks, or contributing to research and innovation in the field of robotics. So, embrace the assembly process with enthusiasm and precision, and let your robot take its first steps into the world.

Now that you've successfully assembled your robot platform, it's time to embark on the next phase of your robotic journey: testing and iterating for optimization. This phase is akin to the trials and refinements that shape a sculptor's masterpiece or a chef's signature dish. It's where you put your robot to the test, identify areas for improvement, and fine-tune its performance to achieve your intended goals.

Think of testing as a series of experiments, each designed to evaluate different aspects of your robot's functionality. Start with basic tests to ensure that the core components are functioning as expected. Does the robot move as intended? Do the sensors provide accurate data? Is the microcontroller processing information correctly?

For instance, if you've built a robot for warehouse automation, the initial tests might involve checking if it can navigate a predefined path without collisions and if it can accurately detect and pick up objects.

As you conduct these initial tests, document your observations and any issues that arise. This documentation will serve as a valuable reference as you iterate and refine your robot.

Once you've confirmed that the fundamental functionalities are in order, it's time to move on to more complex tests that align with your robot's intended application. Consider scenarios and challenges that your robot may encounter in its real-world environment.

If your robot is designed for search and rescue missions, test its ability to navigate through cluttered and unpredictable terrain. Measure its response time to identify and reach simulated victims.

Testing can also involve evaluating the robot's performance under different environmental conditions. For example, if you're building an agricultural robot, assess how it handles various soil types, weather conditions, and crop varieties.

Don't forget to consider safety measures during testing. Depending on your robot's application, you may need to implement emergency stop mechanisms or safety protocols to protect both the robot and its operators.

Imagine you're testing a medical robot designed for surgery assistance. Safety measures, such as fail-safes and remote shutdown capabilities, are critical to ensure the well-being of patients and medical staff.

Throughout the testing phase, collect data and analyze the results. This data-driven approach allows you to identify areas where your robot excels and areas that require improvement.

Look for patterns, trends, and anomalies in the data to gain insights into your robot's performance.

For example, if you're testing a robot for package delivery, analyze delivery times, accuracy, and any instances of missed or incorrect deliveries. This analysis informs you about the robot's efficiency and reliability.

As you analyze the data, be open to feedback and input from others, especially if you're working on a collaborative project or within a research team. Different perspectives can uncover valuable insights and potential solutions to challenges you may encounter.

Consider a scenario where you're part of a team developing a robot for educational purposes. Collaborative feedback helps refine the robot's teaching methods and content delivery, enhancing its effectiveness as an educational tool.

Now, let's talk about the iterative process. Based on your testing results and feedback, start making improvements and refinements to your robot. This is where innovation and creativity come into play. Think of each iteration as a step closer to achieving your robot's full potential.

If your robot struggled with navigating tight spaces during testing, you might explore alternative mobility solutions, such as different wheel configurations or even a combination of wheels and legs.

Suppose you're developing a robot for household assistance. Iterations could involve enhancing its object recognition capabilities, improving its natural language processing for communication, and refining its mobility for tasks like fetching objects.

As you implement changes and enhancements, it's essential to keep detailed records of each iteration. Document the modifications made, the reasons behind them, and the impact on the robot's performance.

Why is documentation crucial? It serves as a roadmap of your robot's evolution. It allows you to backtrack if a particular iteration doesn't yield the desired results or if new challenges arise.

Imagine you're part of a team working on a robot for space exploration. Detailed documentation ensures that future missions can build upon the knowledge and improvements made during previous missions.

Testing and iterating also involve addressing unexpected challenges and setbacks. Robotics is a field where unexpected surprises are common. Components may fail, sensors may encounter interference, or environmental conditions may change abruptly.

It's essential to maintain a resilient and problem-solving mindset during these moments. Treat challenges as opportunities to learn and innovate. Seek solutions, whether it involves troubleshooting hardware issues, refining software algorithms, or adjusting the robot's design.

Consider a scenario where you're developing a robot for underwater exploration. Unforeseen equipment malfunctions can be addressed through meticulous problem-solving, ensuring that the robot can continue its mission.

Continuous testing and iteration are ongoing processes that accompany the lifecycle of your robot. Even after your robot is deployed for its intended purpose, it's essential to monitor its performance and gather real-world data.

Real-world data provides valuable insights that may not be apparent during controlled testing. It helps you understand how your robot interacts with its environment, adapts to changing conditions, and performs over extended periods.

If you've created a robot for environmental monitoring, real-world data allows you to track changes in environmental conditions over time, contributing to valuable research and conservation efforts.

Moreover, consider the potential for remote monitoring and control. Advances in robotics technology often enable remote access to your robot's systems, allowing you to make adjustments and updates as needed.

Imagine you've developed a robot for disaster response. Remote monitoring and control allow you to adapt the robot's capabilities in real-time to respond effectively to evolving emergency situations.

In summary, the testing and iteration phase of robotics development is a dynamic and creative process. It involves evaluating your robot's performance, analyzing data, making improvements, and embracing challenges as opportunities for growth. Through thorough testing and continuous refinement, your robot evolves, becoming better equipped to fulfill its intended purpose. So, approach this phase with curiosity and determination, knowing that each iteration brings you closer to achieving your robotic goals.

BOOK 3
ADVANCED TECHNIQUES IN ROBOTICS RESEARCH BECOMING A SPECIALIST

ROB BOTWRIGHT

Chapter 1: Emerging Trends in Robotics Research

Picture a world where robots work alongside healthcare professionals to enhance patient care, perform delicate surgeries with precision, and assist in tasks that demand accuracy and reliability. This is the realm of robotics in healthcare and medical applications, a field that holds immense promise for revolutionizing the way we approach medical treatments and patient support.

In this chapter, we'll explore the fascinating landscape of robotics in healthcare and delve into the myriad ways these technological marvels are making a significant impact on the well-being of individuals worldwide.

To begin, let's take a closer look at the role of robots in surgery. Imagine a surgical robot, guided by a skilled surgeon, performing minimally invasive procedures with unparalleled precision. These robots are equipped with advanced imaging and navigation systems that provide real-time feedback, allowing surgeons to make precise incisions and operate with exceptional accuracy.

Surgical robots are particularly valuable in procedures that require precision and access to intricate anatomical structures. They reduce the invasiveness of surgery, leading to shorter recovery times and less post-operative pain for patients.

Consider a scenario where a patient requires heart surgery. Robotic-assisted surgery allows the surgeon to perform the procedure through small incisions, resulting in less trauma to the chest and a quicker return to normal activities.

In addition to surgical applications, robots play a crucial role in diagnostics and patient monitoring. Imagine a robot equipped with sensors and imaging technology that can conduct thorough medical scans and tests. These robots provide valuable data to healthcare professionals, aiding in the early detection and diagnosis of diseases.

For instance, robots can assist in conducting ultrasounds, MRIs, and CT scans with exceptional precision. They follow pre-programmed protocols to ensure consistent and accurate imaging, reducing the likelihood of human error.

Think about the benefits of early cancer detection. Robots can assist in performing biopsies and imaging scans, enabling healthcare providers to identify tumors at an early, more treatable stage.

Robots are also being used to enhance patient care in rehabilitation settings. Consider a robot-assisted therapy session for a patient recovering from a stroke. These robots guide patients through tailored exercises, providing real-time feedback and tracking progress. They can adjust the level of assistance based on the patient's abilities, making rehabilitation more effective.

Robot-assisted rehabilitation not only improves patient outcomes but also eases the burden on healthcare providers by allowing for more efficient and consistent therapy sessions.

Now, let's explore the world of telemedicine and telepresence robots. Imagine a scenario where a specialist in a different city or country can remotely examine a patient through a telemedicine robot. These robots are equipped with cameras, microphones, and displays, allowing healthcare professionals to interact with patients in real-time.

Telepresence robots are particularly valuable in rural or underserved areas where access to specialized medical care may be limited. They bridge the gap by connecting patients with experts, enabling consultations and diagnoses without the need for extensive travel.

Think about a patient in a remote village who requires the expertise of a neurologist. A telepresence robot allows the neurologist to conduct a virtual examination and provide guidance to local healthcare providers, improving access to specialized care.

Another area where robots are making a significant impact is in the field of prosthetics and assistive devices. Imagine a person with a lower limb amputation using a robotic prosthetic leg that mimics the natural movement of a human leg. These advanced prosthetics offer greater mobility and functionality, allowing individuals to lead more active lives.

Robotic prosthetics use sensors and microprocessors to detect the wearer's movements and adjust the prosthesis accordingly. This technology provides a level of control and comfort that was previously unimaginable.

Consider the life-changing impact for a veteran who lost a leg in combat. A robotic prosthetic leg enables them to walk, run, and engage in physical activities with confidence and independence.

Beyond surgical procedures and diagnostics, robots are also playing a vital role in the pharmaceutical industry. Imagine a robotic system that can accurately dispense medications, reducing the risk of human errors in drug preparation. These robots ensure precise dosages and improve medication management for patients.

Robotic automation in pharmaceuticals extends to drug discovery and development. Robots can perform high-throughput screening of potential drug compounds, significantly accelerating the research process. This allows pharmaceutical companies to bring new treatments to market more quickly.

Think about the potential for developing life-saving medications. Robots can analyze vast datasets and conduct experiments at a pace that was once unimaginable, bringing us closer to breakthroughs in disease treatment.

Robots are even finding their way into the realm of mental health care. Imagine a robot companion designed to provide emotional support and companionship to individuals experiencing loneliness or anxiety. These robots use artificial intelligence to engage in conversations, offer encouragement, and monitor the emotional well-being of their users.

Robot companions can be particularly valuable for older adults who may be isolated or struggling with mental health issues. They provide a sense of companionship and can alert healthcare providers or family members if they detect signs of distress.

The integration of robotics into healthcare isn't without its challenges and considerations. Privacy and data security are paramount, especially in telemedicine and robotics involving patient information. Ensuring that robots adhere to strict privacy protocols is crucial for building trust and maintaining patient confidentiality.

Moreover, there are ethical considerations when it comes to the use of robots in healthcare decision-making. While robots can assist in diagnosis and treatment recommendations, the final decisions should always involve the expertise and judgment of human healthcare providers.

As we navigate the evolving landscape of robotics in healthcare, it's essential to strike a balance between technological advancements and the compassionate, human-centered care that defines the healthcare industry.

In summary, robotics in healthcare and medical applications represents a remarkable frontier in the world of healthcare. These robots have the potential to enhance surgical precision, improve diagnostics, enable telemedicine, revolutionize rehabilitation, and provide invaluable support in various aspects of patient care. As we continue to explore the possibilities, we must prioritize patient well-being, privacy, and ethical considerations to ensure that robots and humans work together harmoniously to advance healthcare and improve lives.

In the ever-evolving landscape of robotics, one of the most captivating and promising frontiers is the realm of human-robot collaboration and assistance. It's a world where robots and humans work together as partners, leveraging each other's strengths to accomplish tasks, solve problems, and enhance various aspects of our lives.

Imagine a future where robots are seamlessly integrated into our daily routines, not as mere tools but as intelligent companions that anticipate our needs and assist us in a wide range of activities. This chapter explores the multifaceted facets of human-robot collaboration and assistance, shedding light on the ways in which these interactions are reshaping industries, augmenting our capabilities, and redefining the boundaries of what's possible.

To begin our journey, let's delve into the concept of collaborative robotics. Picture a factory floor where human workers operate alongside robotic counterparts. These robots are not isolated behind safety barriers but share the same workspace as their human colleagues, working in close proximity.

Collaborative robots, often referred to as "cobots," are designed to be safe, adaptable, and user-friendly. They can take on repetitive, physically demanding tasks, allowing human workers to focus on more complex and creative aspects of their jobs.

Consider a scenario in an automotive assembly plant. Human workers and cobots collaborate in assembling car components, with robots handling heavy lifting and precision tasks, while humans oversee the process and make crucial decisions.

In healthcare, robots are making a significant impact by assisting medical professionals in various capacities. Imagine a surgical team with a robot that serves as a surgical assistant, precisely passing surgical instruments to the surgeon and providing real-time information on the patient's condition.

Surgical robots are designed to work hand-in-hand with human surgeons, enhancing their capabilities and enabling minimally invasive procedures with unparalleled precision. This collaborative approach leads to better patient outcomes and shorter recovery times.

Now, let's explore the role of robots in education and learning assistance. Imagine a robot tutor that supports students in their studies, offering personalized guidance and feedback. These robots adapt their teaching methods to the individual needs and learning styles of each student. Robot tutors can provide additional support in subjects where students may struggle, reinforcing concepts and providing practice exercises. They offer a unique blend of patience, adaptability, and accessibility that can be particularly valuable in educational settings.

Think about a student who is struggling with mathematics. A robot tutor can patiently explain concepts, offer interactive examples, and track the student's progress, providing targeted assistance where it's needed most.

Robots are also stepping into the role of companions for the elderly. In an aging population, social isolation can be a significant concern. Imagine a robot companion that engages in conversations, plays games, and provides companionship to older adults.

These companionship robots are equipped with artificial intelligence that enables natural language processing, facial recognition, and emotional understanding. They can provide emotional support and help alleviate feelings of loneliness.

Consider the impact on an older adult living alone. A companion robot can provide not only social interaction but also reminders for medication, appointments, and safety checks. Now, let's explore the realm of search and rescue operations. Picture a disaster-stricken area where robots collaborate with rescue teams to locate and assist survivors. These robots are equipped with sensors, cameras, and mobility capabilities that allow them to navigate hazardous environments.

Search and rescue robots can enter collapsed buildings, explore rubble, and provide critical data to rescue teams. They can locate survivors, assess their condition, and even deliver essential supplies.

Consider a natural disaster scenario, such as an earthquake. Search and rescue robots can access areas that are too dangerous for human rescuers, increasing the likelihood of saving lives.

The world of retail is also experiencing a transformation through the use of robots. Imagine a retail store with autonomous robots that manage inventory, restock shelves, and assist customers. These robots enhance operational efficiency and customer service.

Retail robots can scan barcodes, monitor inventory levels, and even provide information to shoppers. They ensure that products are readily available and assist customers in finding items within the store. Think about the convenience for shoppers who can rely on robots to quickly locate products, check prices, and provide recommendations based on their preferences.

Beyond these applications, robots are making strides in providing assistance to individuals with disabilities. Consider a person with limited mobility who can control a robotic exoskeleton, allowing them to stand, walk, and regain a level of independence.

Robotic exoskeletons use sensors and actuators to detect the user's movements and provide the necessary support. They are valuable tools for rehabilitation and enhancing the quality of life for individuals with mobility challenges.

Imagine the life-changing impact for someone who regains the ability to walk with the assistance of a robotic exoskeleton, regaining mobility and independence.

The field of human-robot collaboration and assistance is marked by its potential to transform industries, improve the quality of life, and address complex challenges. However, it also raises ethical considerations, such as privacy, safety, and the potential for job displacement. Privacy is a paramount concern when robots are integrated into our homes, workplaces, and healthcare settings. Ensuring that robots respect personal boundaries and data security is essential for building trust.

Safety is another critical consideration, particularly in collaborative settings. Robots must be equipped with advanced sensors and algorithms to detect and respond to unexpected situations to prevent accidents.

The potential for job displacement is a topic of ongoing debate. While robots can handle repetitive tasks, there is a growing need for individuals who can design, operate, and maintain these robots. This shift may require a focus on reskilling and upskilling the workforce.

In summary, human-robot collaboration and assistance represent a dynamic and transformative frontier in robotics. These interactions between humans and robots have the power to revolutionize industries, improve healthcare, enhance education, support the elderly, aid in disaster response, and redefine the retail experience. As we navigate this evolving landscape, it's essential to address ethical considerations, prioritize safety and privacy, and ensure that technology serves as a tool for enhancing human capabilities and well-being.

Chapter 2: Advanced Kinematics and Dynamics

In the realm of robotics, the study of dynamic analysis holds a crucial place as it unravels the intricate movements and behaviors of robotic manipulators. These mechanisms, often referred to as robotic arms, are the workhorses behind many robotic applications, from industrial automation to healthcare and beyond. In this chapter, we embark on a journey to explore the dynamic analysis of robotic manipulators, unveiling the principles and significance that underpin their operation.

Let's begin by painting a picture of a robotic manipulator. Imagine a mechanical arm composed of multiple segments, each connected by joints. These joints allow the arm to articulate and move in a coordinated manner, much like the flexibility of the human arm. Robotic manipulators are designed to mimic human dexterity, albeit with precision and repeatability that surpass human capabilities.

Consider a manufacturing assembly line where robotic arms diligently assemble intricate components with speed and accuracy. Dynamic analysis plays a pivotal role in ensuring that these robotic arms perform tasks with the desired precision, manipulate objects effectively, and respond to changing conditions in real-time.

Now, let's dive into the essence of dynamic analysis. At its core, dynamic analysis is the study of how forces and torques propagate through a robotic manipulator as it moves. It delves into the interplay of kinematics (the study of motion) and kinetics (the study of forces and torques) to understand how a robotic arm responds to external loads, accelerations, and disturbances.

Imagine a robotic arm lifting a heavy object. Dynamic analysis allows us to comprehend the intricate dance of forces within the arm's joints and segments. It helps us predict how the arm will respond to the weight, how the joints will exert torques to counteract gravity, and how the arm's motion will evolve as it lifts and moves the object.

Dynamic analysis relies on mathematical models to represent the complex relationships between forces, torques, and motion. These models, often expressed in the form of equations, take into account the physical properties of the robotic manipulator, including its mass distribution, moments of inertia, and friction. They consider the kinematic chain of the arm, detailing how each joint's motion affects the others.

Consider the mathematical elegance of these models. Equations of motion capture the dynamic behavior of each joint and segment, providing insights into the forces required to achieve a desired motion. These equations consider the accelerations, velocities, and positions of each component, making it possible to predict and control the arm's behavior.

Now, let's explore the practical implications of dynamic analysis in robotics. Imagine a scenario where a robotic manipulator is tasked with painting a complex pattern on an irregular surface. Dynamic analysis helps in optimizing the arm's movements, ensuring that it applies just the right amount of force and follows the intended trajectory.

Dynamic analysis also aids in the development of advanced control algorithms. Imagine a robot performing surgery with millimeter precision. Dynamic models enable the design of control strategies that compensate for external forces, ensuring that the robot can adapt to the surgeon's movements while maintaining the utmost precision.

Consider a scenario in space exploration, where robotic arms are deployed on rovers to manipulate scientific instruments and collect samples. Dynamic analysis is indispensable in designing these arms to withstand the harsh conditions of space, including microgravity and extreme temperatures.

Dynamic analysis doesn't stop at just understanding the forces and torques within robotic manipulators. It extends to the realm of trajectory planning and optimization. Imagine a robot tasked with welding intricate seams on an automotive assembly line. Dynamic analysis informs the planning process, ensuring that the arm follows the most efficient and stable path while applying the correct forces to achieve flawless welds.

Dynamic analysis is also vital in predicting the wear and tear on robotic components. Consider a manufacturing facility with robotic arms tirelessly performing tasks around the clock. Dynamic analysis allows engineers to estimate the lifespan of critical components, such as gears and bearings, by assessing the forces they endure during operation.

Imagine a scenario in agriculture, where robotic arms are used for precision harvesting. Dynamic analysis guides the design of these arms to minimize soil compaction while efficiently plucking ripe fruits from the trees, thus optimizing the agricultural yield and sustainability.

In essence, dynamic analysis serves as a compass guiding the development, operation, and optimization of robotic manipulators across a wide range of industries. Its applications span from industrial automation to healthcare, from space exploration to agriculture, and from manufacturing to artistry.

However, delving into the world of dynamic analysis is not without its challenges. It involves dealing with complex mathematical models and computational simulations, demanding a deep understanding of both mechanical engineering and mathematics.

Consider the complexities of modeling friction and backlash in robotic joints. Dynamic analysis requires engineers to account for these nonlinear effects, ensuring that the models accurately reflect the real-world behavior of robotic manipulators.

Moreover, dynamic analysis often necessitates high-performance computing resources to solve the intricate equations of motion in real-time. Imagine a scenario in autonomous vehicles where dynamic analysis is essential for controlling the vehicle's robotic arm. Real-time computation is critical to ensuring the arm responds swiftly and accurately to changing conditions on the road.

Within the intricate world of robotics, there lies a fascinating challenge that both engineers and robots must navigate: the concept of singularities and the art of redundancy resolution. These topics are crucial to the operation of robotic systems, influencing their ability to perform complex tasks with precision and agility.

Let's embark on a journey to understand singularities and redundancy resolution in the realm of robotics, unravelling their significance and practical implications. Imagine a robotic arm, sleek and graceful, capable of executing a wide range of movements. This arm consists of multiple segments, linked together by joints that grant it flexibility and mobility.

Now, picture a scenario where the robotic arm is tasked with reaching a specific point in space. Singularities are critical moments that can occur during such movements. A singularity is a configuration of the robotic arm where it loses degrees of freedom, making certain movements impossible or highly challenging to execute.

Consider a scenario where a robotic arm is reaching for an object on a table. At a singularity point, the arm's joints align in such a way that a small change in joint angles could result in a significant change in the arm's position. This can lead to unpredictability and difficulty in accurately controlling the arm's movement.

Singularities are not limited to a single point but can occur within a region of the arm's workspace. Imagine the robotic arm painting a car's surface. As it moves across the car's curved body, it encounters multiple singularities where the arm's dexterity is compromised. These moments require careful planning and control to ensure smooth and precise painting.

To navigate singularities effectively, engineers rely on redundancy resolution techniques. Redundancy, in the context of robotics, refers to the presence of more degrees of freedom than are necessary to perform a task. Think of it as having extra joints in the robotic arm that can be used to optimize performance and avoid singularities.

Imagine a scenario where a robotic arm is used for pick-and-place operations in a warehouse. Redundancy allows the arm to choose from multiple joint configurations to reach a target object while avoiding singularities. It can optimize the path to minimize energy consumption or maximize speed, depending on the task.

Consider a surgical robot assisting a surgeon during a delicate procedure. Redundancy resolution enables the robot to adjust its joint angles to avoid singularities while maintaining precise control. It ensures that the robot can adapt to the surgeon's movements and provide stable support.

Now, let's explore the mathematics behind singularities and redundancy resolution. Imagine a robotic arm with six joints. The position and orientation of the arm can be described using a mathematical representation known as the transformation matrix. Singularities occur when the matrix becomes singular, meaning it cannot be inverted to calculate joint angles uniquely.

Mathematically, singularities are identified by a determinant of the transformation matrix becoming zero. Engineers use algorithms and numerical techniques to detect and avoid singularities during robotic operations. This involves analyzing the arm's configuration and planning movements to steer clear of singularity regions.

Consider a scenario where a robot is used for 3D printing complex structures. Redundancy resolution algorithms can optimize the robot's joint angles to ensure even deposition of material while avoiding singularities. This results in precise and efficient 3D printing.

Now, let's dive into the practical applications of redundancy resolution. Imagine a scenario in aerospace manufacturing, where robots are used to assemble intricate components. Redundancy resolution allows the robots to adapt to varying component shapes and sizes, optimizing their movements for each task.

Consider a scenario in the film industry, where robotic cameras are used to capture dynamic shots. Redundancy resolution enables the cameras to smoothly follow actors and objects, adjusting their angles and positions in real-time to achieve cinematic perfection.

In the field of rehabilitation, robotic exoskeletons assist individuals with mobility impairments. Redundancy resolution algorithms ensure that the exoskeletons adapt to the wearer's movements and provide the necessary support, making it possible for people to regain mobility.

Imagine a scenario in underwater exploration, where robotic arms are used to manipulate tools and collect samples on the ocean floor. Redundancy resolution techniques allow the arms to adjust to the underwater environment and perform tasks with precision, even in challenging conditions.

Singularities and redundancy resolution are not limited to industrial applications. Consider a scenario in entertainment, where animatronic characters delight audiences at theme parks. Redundancy resolution ensures that these characters move smoothly and convincingly, creating immersive and magical experiences.

Now, let's address the challenges posed by singularities and the complexity of redundancy resolution. Singularities can be unpredictable, making it challenging to plan robotic movements in advance. Engineers must develop robust algorithms and control strategies to handle singularities effectively.

Consider a scenario in autonomous vehicles, where robotic arms are used for maintenance and repair tasks. Redundancy resolution is essential to ensure that the arms can adapt to different vehicle models and repair scenarios. It requires advanced sensing and planning capabilities.

In the world of human-robot collaboration, redundancy resolution becomes a critical factor. Imagine a scenario in a manufacturing facility where robots work alongside human workers. Redundancy resolution algorithms allow the robots to adjust their movements to accommodate human colleagues and avoid collisions.

Chapter 3: Control Strategies for Complex Robotic Systems

In the world of robotics, model-based control techniques serve as the guiding compass, enabling robots to navigate complex environments and execute precise movements with grace and efficiency. These techniques are the backbone of robotic control systems, providing the means to transform mathematical models into real-world actions.

Imagine a scenario where a robotic arm must delicately pick up a fragile object, such as a glass vase, without shattering it. Model-based control techniques come into play to orchestrate the arm's movements with utmost precision. These techniques involve the development of mathematical models that capture the dynamics, kinematics, and physics of the robotic system.

Now, let's delve deeper into the essence of model-based control techniques. At their core, these techniques involve the creation of a mathematical representation of the robot's behavior. This representation encapsulates the relationships between the robot's inputs, such as joint angles and forces, and its outputs, which include positions, velocities, and torques.

Consider a scenario in which a robotic vehicle must navigate a rough terrain with potholes and obstacles. Model-based control techniques allow the vehicle to anticipate the effects of terrain variations on its motion. By incorporating information about the vehicle's suspension, wheel dynamics, and terrain properties into the mathematical model, the control system can adjust wheel torques in real-time to ensure a smooth and stable ride.

Imagine a scenario in which a humanoid robot is designed to walk with human-like grace and fluidity. Model-based control techniques play a pivotal role in creating a dynamic model of the robot's body, capturing the interplay of joint angles, muscle forces, and the resulting motion. This model guides the control system in orchestrating the robot's movements, ensuring that it maintains balance and adaptability in different environments.

Now, let's explore the practical applications of model-based control techniques. Imagine a scenario in which a robotic surgical assistant is used to perform delicate procedures with unmatched precision. Model-based control techniques enable the robot to mimic the movements of a skilled surgeon, translating the surgeon's inputs into precise and steady instrument movements within the patient's body.

Consider a scenario in which a drone is tasked with inspecting a complex structure, such as a bridge or a high-rise building. Model-based control techniques allow the drone to account for aerodynamic forces, wind disturbances, and the geometry of the structure. This enables the drone to autonomously navigate close to the structure, capturing detailed inspections and data without collisions.

In the realm of autonomous driving, model-based control techniques are indispensable. Imagine a self-driving car navigating through urban traffic. The control system relies on a dynamic model of the car's motion, taking into account factors such as wheel dynamics, tire-road interactions, and vehicle dynamics. This model guides the car in making real-time decisions for safe and efficient driving.

Now, let's dive into the mathematical intricacies of model-based control techniques. These techniques often involve differential equations that describe the evolution of the robot's state variables over time. These state variables encompass a wide range of information, including positions, velocities, accelerations, and forces.

Consider a scenario in which a robotic manipulator is used for material handling in a manufacturing facility. Model-based control techniques require the development of a dynamic model that characterizes the manipulator's behavior as it moves to grasp, lift, and transport objects. This model forms the foundation for the control system's ability to plan and execute precise movements.

Imagine a scenario in which a quadcopter drone must hover steadily in the presence of gusty winds. Model-based control techniques leverage a dynamic model of the drone's aerodynamics, accounting for the forces acting on its rotors, the drone's mass distribution, and external disturbances. This model guides the control system in adjusting rotor speeds to maintain stable hover.

Now, let's address the challenges posed by model-based control techniques. While these techniques offer precision and adaptability, they also require accurate modeling of the robot and its environment. Inaccuracies in the mathematical models can lead to discrepancies between the desired and actual robot behavior.

Consider a scenario in which a robot is tasked with assembling intricate electronic components. Model-based control techniques demand a high level of accuracy in modeling the robot's end-effector, the components' geometry, and the assembly process. Any modeling errors can result in misalignment or damage to the components.

In the field of legged robotics, such as quadrupedal robots, achieving stable locomotion through model-based control is a formidable challenge. These robots must account for changing terrain, leg dynamics, and body stability. Model-based control techniques require precise dynamic models to ensure coordinated and stable movements.

In summary, model-based control techniques are the cornerstone of robotic control systems, enabling robots to perform tasks with precision and adaptability across a wide range of applications. These techniques involve the development of mathematical models that capture the intricacies of robotic behavior, from manipulators to vehicles to drones. While they offer immense potential, they also demand careful modeling and calibration to achieve the desired level of accuracy and reliability. As robotics continues to advance, the mastery of model-based control techniques will be pivotal in shaping the future of automation and technology. In the intricate realm of robotics, where precise and dynamic control is paramount, nonlinear control approaches emerge as powerful tools to navigate the complexities of robotic systems. These approaches offer the means to handle highly dynamic, nonlinear, and unpredictable environments, making them indispensable for a wide array of robotic applications.

Imagine a scenario where a quadrupedal robot is traversing rough terrain, scaling uneven surfaces, and maintaining balance in the face of unpredictable disturbances. Nonlinear control approaches come into play, allowing the robot to adapt its gait and joint movements in real-time to maintain stability and continue its mission.

Now, let's delve deeper into the essence of nonlinear control approaches. At their core, these approaches recognize and embrace the inherent nonlinearities present in robotic systems. Unlike linear control methods, which assume a linear relationship between inputs and outputs, nonlinear control acknowledges that real-world systems are often governed by intricate, nonlinear equations.

Consider a scenario where a drone is performing aerial acrobatics, executing flips and rolls with precision. Nonlinear control approaches account for the complex aerodynamic forces and moments at play during these maneuvers. They leverage mathematical models that capture the drone's nonlinear dynamics, enabling it to execute intricate maneuvers while maintaining control.

Imagine a scenario in the field of soft robotics, where robots are designed with flexible, deformable bodies. Nonlinear control approaches are instrumental in orchestrating the movements of these robots, as they often involve intricate interactions between material deformation, forces, and environmental conditions. These approaches allow soft robots to adapt to their surroundings and execute tasks with grace and finesse. Now, let's explore the practical applications of nonlinear control approaches. Consider a scenario in which a robotic exoskeleton is aiding a person with mobility impairments to walk. Nonlinear control techniques enable the exoskeleton to adapt to the wearer's movements and provide the necessary support. They account for the nonlinear relationship between joint angles, muscle forces, and human motion. Imagine a scenario in which a swarm of autonomous underwater vehicles (AUVs) is tasked with exploring the depths of the ocean. Nonlinear control approaches are essential for coordinating the movements of these AUVs in the presence of underwater currents, pressure changes, and varying terrain. They enable the AUVs to maintain formation and efficiently explore the underwater environment.

In the context of humanoid robots, which mimic the complexities of human movement, nonlinear control approaches play a pivotal role. Imagine a scenario where a humanoid robot is dancing with fluidity and grace. Nonlinear control techniques allow the robot to execute intricate dance moves by accounting for the nonlinearities in its joint dynamics and motion patterns. Now, let's dive into the mathematical intricacies of nonlinear control approaches. These approaches often rely on advanced mathematical tools, such as differential equations and optimization techniques, to model and control robotic systems. Differential equations capture the dynamic behavior of robots, describing how variables like positions, velocities, and forces evolve over time.

Consider a scenario in which a robotic arm is used for precise and dexterous tasks, such as assembling intricate electronics. Nonlinear control approaches require the development of a dynamic model that accurately represents the arm's behavior, including the complex interactions between joints, links, and end-effectors. This model serves as the foundation for control algorithms that enable the arm to manipulate objects with precision. Imagine a scenario in which a flying robot, such as a quadcopter, is performing agile maneuvers in a cluttered environment. Nonlinear control techniques leverage optimization algorithms to compute the optimal control inputs in real-time, considering the nonlinear dynamics of the quadcopter and the obstacles in its path. This allows the quadcopter to navigate tight spaces and avoid collisions.

Now, let's address the challenges posed by nonlinear control approaches. While they offer great flexibility and adaptability, they also demand a deep understanding of the robot's dynamics and the development of accurate mathematical models. Inaccuracies in modeling can lead to control instability and suboptimal performance.

Consider a scenario in which a robotic leg prosthesis aims to provide natural and intuitive movement to an amputee. Nonlinear control approaches require precise modeling of the human biomechanics and the prosthesis dynamics to achieve a seamless integration between the amputee's movements and the robotic limb. In the field of autonomous vehicles, where safety is paramount, nonlinear control approaches must be rigorously tested and validated. Imagine a self-driving car navigating complex urban traffic. Nonlinear control techniques must ensure that the car can handle unexpected scenarios, such as sudden braking or swerving, while maintaining safety and avoiding accidents.

In summary, nonlinear control approaches stand as a cornerstone of robotic control, enabling robots to tackle nonlinear dynamics and intricate environments with finesse and adaptability. These approaches embrace the inherent complexities of robotic systems, leveraging advanced mathematics and optimization techniques to model and control robots effectively. As robotics continues to advance, nonlinear control approaches will play an increasingly vital role in shaping the future of automation and technology.

Chapter 4: Localization and Mapping in Robotics

In the world of robotics, navigating through complex environments with precision is a challenge that requires innovative solutions, and one such solution is visual odometry. Visual odometry is the art of using cameras to estimate the motion of a robot by tracking visual features in its surroundings. It's like giving a robot the ability to see and understand its movement, much like how we rely on our eyes to gauge our own motion.

Imagine a scenario where a mobile robot is exploring an unknown environment, like a search and rescue mission in a disaster-stricken area. Visual odometry plays a crucial role in helping the robot make sense of its surroundings. It continuously captures images of the environment, identifies key visual features, and calculates how the robot is moving relative to those features.

Now, let's delve deeper into the essence of visual odometry. At its core, visual odometry relies on a process known as feature tracking or feature matching. This involves identifying distinctive points or landmarks in the images, such as corners, edges, or unique patterns, and then tracking how these features move as the robot changes its position.

Consider a scenario where a drone is flying through a dense forest to locate a lost hiker. Visual odometry becomes the drone's eyes, enabling it to maintain a sense of its position and orientation even when GPS signals are unreliable due to the dense canopy. By tracking the unique visual features in the forest, the drone can estimate its own path and navigate safely.

Imagine a scenario where a self-driving car is maneuvering through a bustling cityscape filled with pedestrians, other vehicles, and complex intersections. Visual odometry, combined with other sensor data, helps the car maintain its lane, avoid obstacles, and make timely decisions. It provides an additional layer of perception, enhancing the car's ability to navigate urban environments.

Now, let's explore the practical applications of visual odometry. Consider a scenario in the world of robotics research, where a robot is designed to autonomously map and explore underground caves. Visual odometry serves as the robot's virtual guide, allowing it to construct detailed maps of the cave system by continuously tracking visual features on the cave walls and floor.

Imagine a scenario in the agricultural industry, where an autonomous tractor is tasked with planting crops in precise rows. Visual odometry enables the tractor to maintain accurate spacing between the rows, ensuring efficient planting. It relies on the visual cues of the planted crops to gauge its own movement.

In the context of augmented reality, visual odometry is employed to anchor virtual objects in the real world. Imagine a scenario where you're using an augmented reality headset to explore a museum. Visual odometry tracks the features in the museum's surroundings, allowing the headset to place virtual exhibits seamlessly within your field of view, creating an immersive experience.

Consider a scenario in the world of robotics competitions, where teams of students build robots to navigate challenging environments. Visual odometry becomes a key tool for these young engineers, helping their robots stay on course and complete tasks accurately. It's a skill they develop and refine as they compete.

Now, let's dive into the technical details of visual odometry. To estimate motion accurately, it relies on complex algorithms that calculate the transformation between consecutive camera frames. These algorithms consider factors such as camera calibration, image distortion correction, and feature correspondence.

Imagine a scenario in which a robotic vacuum cleaner is navigating your home, avoiding furniture and obstacles. Visual odometry algorithms analyze the images captured by the robot's camera and compare them to previous images to determine how far and in which direction the robot has moved. This allows the vacuum cleaner to clean your home efficiently while avoiding collisions.

In outdoor scenarios, where lighting conditions can change dramatically, visual odometry faces challenges. Imagine a scenario where a drone is surveying a construction site from dawn to dusk. As the sun's position shifts, the lighting on the construction site changes, affecting the appearance of visual features. Robust visual odometry algorithms must adapt to these variations to provide accurate motion estimates.

Now, let's address the limitations and challenges of visual odometry. It relies heavily on the quality of visual features in the environment. In scenarios with featureless or repetitive surfaces, such as open water or featureless desert terrain, visual odometry may struggle to provide accurate motion estimates. Consider a scenario in which a robot is exploring the surface of another planet, such as Mars. Visual odometry becomes a critical tool for planetary rovers, but the harsh environmental conditions and the limited availability of distinctive visual features on Mars' surface pose significant challenges. Engineers and scientists continuously work to overcome these obstacles.

In real-time applications, such as autonomous vehicles, visual odometry must operate with low latency to ensure safety and responsiveness. Imagine a self-driving car navigating busy city streets. Visual odometry algorithms must process images quickly and accurately to provide the car with real-time feedback on its position and motion.

In summary, visual odometry stands as a remarkable technology that empowers robots and autonomous systems with the ability to perceive their own motion by analyzing visual information from their surroundings. Whether it's a drone navigating a forest, a self-driving car navigating a city, or a robot exploring caves, visual odometry plays a pivotal role in enhancing their capabilities. As technology continues to advance, the fusion of visual perception and intelligent algorithms will further elevate the role of visual odometry in robotics and automation, shaping the future of navigation and exploration. Navigating the world successfully is a monumental challenge for robots, especially when they operate in diverse environments, ranging from bustling urban landscapes to remote outdoor terrains. To overcome this challenge, robots rely on a combination of sensors and sophisticated algorithms to understand their surroundings and determine their position accurately. Among the essential components of this navigation toolkit is map fusion and global localization, which play a pivotal role in helping robots navigate and operate effectively in the real world.

Imagine a scenario where a delivery robot is making its way through a complex city, filled with narrow alleys, crowded streets, and a myriad of potential obstacles. Map fusion and global localization are the guiding lights for this robot, enabling it to seamlessly merge data from various sensors and maps, such as lidar, cameras, and pre-existing maps, to create a comprehensive representation of its environment.

Intriguingly, map fusion does not just involve combining sensor data and existing maps; it also includes the fusion of maps from multiple sources. Picture a scenario where an autonomous vehicle is driving through a city center. It can benefit from fusing data not only from its onboard sensors but also from other vehicles in its vicinity, creating a shared map that enhances the overall awareness and safety of all vehicles involved.

Global localization is the process of determining a robot's precise position within a known global reference frame. Think of it as a GPS system for robots, but with greater precision and adaptability. In a world where GPS signals can be unreliable or unavailable, global localization becomes paramount. It allows a robot to know not only where it is but also how confident it is about that location.

Consider a scenario where a search and rescue robot is exploring a disaster-stricken area with collapsed buildings and debris. Global localization provides the robot with the ability to pinpoint its exact location within the disaster site, enabling it to communicate its findings with rescue teams and efficiently navigate the treacherous terrain.

Now, let's dive into the technical aspects of map fusion and global localization. Map fusion involves aligning and combining data from different sources, ensuring that the resulting map accurately reflects the robot's environment. This process often includes addressing sensor calibration, correcting for sensor inaccuracies, and dealing with uncertainties in the data.

Imagine a scenario where a robot is exploring the depths of the ocean, conducting underwater research. Map fusion allows the robot to merge data from its underwater sonar sensors with previously collected maps, creating a comprehensive underwater map that can aid in scientific exploration and mapping of the ocean floor.

Global localization relies on a combination of sensor data and map information. To determine its global position, a robot often employs techniques such as particle filters or extended Kalman filters. These algorithms take into account the robot's sensor measurements, the map data, and the robot's motion model to estimate its position accurately.

Consider a scenario where an agricultural robot is autonomously tending to crops in a vast field. Global localization ensures that the robot can navigate the field accurately, even in the absence of GPS signals. It relies on its onboard sensors and pre-existing maps of the field to determine its position, allowing it to perform tasks such as precision planting and harvesting.

Now, let's explore some of the challenges and complexities associated with map fusion and global localization. One significant challenge is dealing with sensor inaccuracies and uncertainties. Sensors can provide noisy or incomplete data, which can lead to errors in the fused maps and localization estimates. Engineers and researchers continually work on improving sensor technologies and developing robust algorithms to address these challenges.

Imagine a scenario where a robot is exploring a dense forest, searching for wildlife. Map fusion becomes a critical tool in helping the robot build an accurate map of the forest, but the presence of dense foliage, variable lighting conditions, and dynamic terrain can make the task challenging. Robust algorithms are essential to handle these complexities.

Another challenge arises when robots need to operate in dynamic environments where the surroundings change over time. Consider a scenario where an autonomous drone is monitoring a construction site. Map fusion and global localization must adapt to the evolving nature of the site, accounting for newly erected structures, moving equipment, and changing terrain.

In summary, map fusion and global localization are integral components of robotic navigation and perception. They enable robots to understand their surroundings, create accurate maps, and determine their precise position within a global reference frame. Whether it's a delivery robot navigating city streets, an underwater explorer mapping the ocean floor, or an agricultural robot tending to crops, these technologies empower robots to operate effectively in diverse and dynamic environments. As technology advances, map fusion and global localization will continue to play a crucial role in shaping the future of robotics, paving the way for more capable and adaptable robotic systems.

Chapter 5: Computer Vision and Object Recognition

Welcome to the fascinating world of image processing for robotic vision. In this chapter, we'll explore how robots perceive and understand their environment through the lens of cameras and the power of image processing techniques. Imagine a robot equipped with cameras, whether it's an autonomous car navigating busy streets or a robotic arm picking items off a conveyor belt in a factory. These robots rely on image processing to make sense of the visual information they capture and make informed decisions.

At the core of image processing for robotic vision is the ability to extract valuable information from images. Think of it as teaching robots to see and interpret the world around them, much like how we humans use our eyes to understand our surroundings. However, unlike human vision, which is instinctive and effortless, robotic vision requires careful programming and sophisticated algorithms.

Consider a scenario where a surveillance robot is patrolling a large warehouse. Its cameras capture images of the facility, and image processing algorithms analyze these images to detect any suspicious activities. These algorithms can identify people, objects, and movements, providing the robot with the necessary information to alert security personnel if needed.

One fundamental aspect of image processing is image filtering. Imagine a scenario where a robot is exploring a polluted river to assess water quality. Image filtering techniques can be used to enhance the visibility of contaminants in the water, making it easier for the robot to detect and analyze pollution levels. These techniques can also help in noise reduction, improving the overall quality of the images.

Another crucial concept in image processing is image segmentation. Think of it as teaching a robot to recognize and separate different objects or regions within an image. In the context of autonomous vehicles, image segmentation plays a pivotal role in identifying lane markings, traffic signs, and other vehicles on the road. This information is vital for the vehicle's navigation and safety systems.

Imagine a scenario where a robot is deployed in a warehouse to assist with inventory management. Image segmentation techniques can help the robot distinguish between different products on the shelves, enabling it to count items accurately and track inventory levels. This not only saves time but also reduces errors in inventory management.

Robotic vision also relies heavily on object detection and recognition. Robots must not only identify objects but also understand what those objects are and their relevance to the task at hand. Consider a scenario where a robotic assistant is helping in a kitchen. Through object detection and recognition, the robot can distinguish between ingredients, utensils, and cooking equipment, allowing it to assist in meal preparation effectively.

Intriguingly, robots can also be trained to recognize and respond to human gestures and expressions through image processing. Imagine a scenario where a robot is designed to assist people with disabilities. By analyzing facial expressions and gestures, the robot can understand a person's emotions and intentions, providing appropriate assistance and support.

Now, let's delve into the technical aspects of image processing. One common technique is edge detection, which involves identifying the boundaries of objects within an image. This is akin to outlining objects to make them stand out, similar to how we perceive the edges of objects in our visual field. Edge detection is crucial for tasks like obstacle avoidance in robotics.

Imagine a scenario where a robot is navigating a cluttered environment. Edge detection algorithms help the robot identify obstacles and plan its path accordingly to avoid collisions.

These algorithms analyze the edges of objects in the robot's field of view and calculate the safest route.

Image registration is another critical concept in image processing for robotic vision. This technique involves aligning multiple images to create a composite view or to track changes over time. Consider a scenario where a drone is conducting aerial surveys of a construction site. Image registration allows the drone to stitch together multiple images into a seamless map, enabling accurate monitoring of construction progress.

In addition to image processing, machine learning plays a significant role in enhancing robotic vision capabilities. Machine learning algorithms can be trained to recognize specific objects, patterns, or even anomalies in images. This learning process allows robots to adapt and improve their visual understanding over time.

Imagine a scenario where a robot is inspecting manufactured products on an assembly line. Machine learning algorithms can be trained to identify defects, such as cracks or defects in the products, ensuring that only high-quality items pass inspection. This level of automation and precision is crucial in manufacturing quality control.

In summary, image processing for robotic vision is a powerful and evolving field that enables robots to see, interpret, and interact with the world around them. Whether it's autonomous vehicles navigating complex traffic, surveillance robots patrolling secure facilities, or robotic assistants assisting in daily tasks, image processing techniques empower robots to make informed decisions based on visual information. As technology continues to advance, we can expect robotic vision to become even more sophisticated, paving the way for safer, more capable, and more intuitive robotic systems.

Deep learning has ushered in a revolutionary era for object detection and recognition in the field of robotics. Picture a world where robots can not only perceive their environment but also identify and understand the objects within it, much like we do as humans. This remarkable capability is made possible through deep learning techniques, particularly Convolutional Neural Networks (CNNs).

Consider a scenario where an autonomous delivery robot navigates a bustling urban environment. Deep learning models, such as CNNs, allow the robot to detect pedestrians, vehicles, traffic signs, and even specific landmarks with remarkable accuracy. These models process visual data from onboard cameras in real-time, providing crucial information for safe navigation.

One of the key strengths of deep learning is its ability to handle complex and unstructured data. In the context of object detection and recognition, this means that robots can identify objects in various shapes, sizes, and orientations, even when they are partially obscured or in cluttered environments.

Imagine a scenario where a robotic assistant is helping with household chores. Deep learning models enable the robot to recognize a wide range of objects, from dishes and cutlery to clothing items and household appliances. This recognition capability allows the robot to interact with objects effectively, whether it's setting the table or doing the laundry.

Deep learning is also instrumental in the development of advanced driver assistance systems (ADAS) for autonomous vehicles. These systems rely on object detection and recognition to ensure the safety of passengers and pedestrians alike. Whether it's identifying other vehicles on the road, recognizing traffic lights, or detecting potential obstacles, deep learning models play a pivotal role in enabling autonomous driving.

Now, let's delve into the technical aspects of deep learning for object detection and recognition. Convolutional Neural Networks, or CNNs, are at the heart of this technology. These neural networks are inspired by the human visual system, with layers of interconnected neurons that can recognize patterns and features in images.

A crucial concept within CNNs is the convolutional layer, which applies a series of filters or kernels to input images. These filters extract specific features from the images, such as edges, textures, or more complex patterns. Imagine a scenario where a robot is inspecting manufactured products on an assembly line. CNNs can be trained to detect defects by recognizing irregularities in the products' surfaces, ensuring high-quality manufacturing.

One of the remarkable properties of CNNs is their ability to learn hierarchical representations of features. This means that as you move deeper into the network's layers, it can recognize increasingly abstract and complex features. In the context of object recognition, this hierarchical approach allows CNNs to understand objects at various levels of detail, from basic shapes and edges to more specific attributes.

Consider a scenario where a surveillance drone is monitoring a large area for security purposes. Deep learning models enable the drone to detect and track objects of interest, such as intruders or suspicious activities. These models can distinguish between humans, animals, and vehicles, providing valuable information to security personnel.

Another critical aspect of deep learning for object detection is the concept of transfer learning. Transfer learning allows models trained on one task to be adapted for another related task with relatively little additional training data. This flexibility is invaluable in robotics, where adapting pre-trained models to specific environments or applications can save time and resources.

Imagine a scenario where a robotic warehouse assistant needs to recognize and sort various products on a conveyor belt. Transfer learning can be employed to adapt a pre-trained object detection model to this specific task, significantly reducing the time required for training and fine-tuning.

An exciting area within deep learning for robotics is the integration of object detection with natural language understanding. This means that robots can not only recognize objects but also understand and respond to human commands related to those objects. For instance, in a smart home setting, a robot can recognize a request to "bring a book" and locate and retrieve the requested item.

In summary, deep learning has revolutionized object detection and recognition in robotics, empowering robots with the ability to perceive and understand their surroundings in unprecedented ways. Whether it's autonomous vehicles navigating complex traffic, robotic assistants helping with daily tasks, or surveillance systems enhancing security, deep learning models, particularly CNNs, have become indispensable tools. As technology continues to advance, we can expect even more sophisticated and capable robots that seamlessly integrate into our lives, thanks to the power of deep learning for object detection and recognition.

Chapter 6: Machine Learning for Autonomous Robots

Reinforcement learning represents a pivotal paradigm in the development of autonomous systems, propelling them towards ever greater levels of adaptability and intelligence. It's a field of artificial intelligence that enables machines to learn from their interactions with the environment, akin to how humans learn from trial and error.

Imagine a scenario where a self-driving car learns to navigate the complexities of city traffic through reinforcement learning. Rather than relying solely on pre-programmed rules, the car continuously assesses its actions and refines its decision-making based on real-world experiences. This learning process allows the car to adapt to diverse and unpredictable situations, ultimately enhancing safety and performance.

At the heart of reinforcement learning lies the concept of an agent, which is the autonomous entity seeking to maximize a cumulative reward by taking actions within an environment. This environment consists of a set of states, actions, and a reward function. The agent interacts with the environment by selecting actions to achieve its objectives while considering the consequences of those actions.

Think of a robot exploring an unfamiliar terrain. The robot is the agent, and the terrain is the environment. At each step, the robot decides whether to move forward, turn, or take other actions. Based on its actions, it receives rewards or penalties— positive rewards for safe movements, negative rewards for encountering obstacles. Over time, the robot learns to choose actions that maximize its overall reward, enabling it to navigate the terrain efficiently.

Reinforcement learning employs a strategy known as the exploration-exploitation dilemma. This balance is essential for an agent to learn effectively. Exploration involves trying out new actions to discover their consequences, even if the outcome is uncertain. Exploitation, on the other hand, entails selecting actions that have previously yielded positive results.

Consider a scenario where a robotic vacuum cleaner uses reinforcement learning to optimize its cleaning efficiency. Initially, it may explore various paths and cleaning patterns, occasionally bumping into obstacles or getting stuck. As it gathers more experience, the vacuum cleaner gradually shifts towards exploitation, focusing on the most effective cleaning strategies it has learned.

Central to reinforcement learning is the notion of a reward signal, which guides the agent's learning process. This signal provides feedback to the agent, indicating how good or bad its actions are in achieving its goals. The agent's objective is to find a policy—a strategy or set of actions—that maximizes the expected cumulative reward over time.

Imagine a scenario where a drone is tasked with delivering medical supplies to remote areas. The reward signal in this case might include factors like delivery speed, accuracy, and safety. The reinforcement learning algorithm helps the drone discover the optimal flight path and actions that maximize the delivery's success while minimizing risks.

Reinforcement learning algorithms come in various forms, including value-based and policy-based methods. Value-based methods aim to estimate the value of taking specific actions in particular states. These methods help the agent decide which actions are most likely to lead to high rewards.

Policy-based methods, on the other hand, focus on learning the policy directly—the mapping from states to actions. This approach is particularly useful in situations where the agent's action space is continuous or high-dimensional.

Consider a scenario where a humanoid robot is learning to walk. Value-based reinforcement learning can be used to estimate the value of different joint movements and body positions. Policy-based methods, on the other hand, can directly learn the motor control policies that guide the robot's movements to maintain balance and achieve stable walking.

One of the defining characteristics of reinforcement learning is its ability to handle complex, non-linear relationships. Deep Reinforcement Learning (DRL) combines reinforcement learning with deep neural networks, enabling agents to tackle high-dimensional state spaces and action spaces.

Think of a scenario where an autonomous drone learns to perform acrobatic maneuvers. Deep reinforcement learning allows the drone to process and analyze sensor data, such as images from onboard cameras, in real-time. It can learn intricate flight patterns and optimize its control actions to execute flips, rolls, and loops with precision.

Another critical aspect of reinforcement learning is the concept of exploration strategies. These strategies determine how the agent explores its environment to gather valuable experience. Common exploration strategies include epsilon-greedy, where the agent sometimes chooses random actions, and optimistic initialization, where the agent starts with optimistic estimates of action values. Imagine a scenario where a robot is learning to play chess. The exploration strategy it employs influences how often it tries different moves during practice games. Over time, the robot refines its strategy, favoring actions that lead to winning positions and gradually reducing the frequency of random moves. Reinforcement learning has far-reaching implications beyond robotics. It has found applications in recommendation systems, natural language processing, and even in optimizing industrial processes. In autonomous systems, reinforcement learning continues to push the boundaries of what machines can achieve.

Consider a scenario where a virtual assistant uses reinforcement learning to personalize content recommendations for users. The assistant learns from user interactions, continuously refining its recommendations to provide a more tailored and engaging experience.

In the realm of autonomous vehicles, reinforcement learning plays a crucial role in improving safety and efficiency. Self-driving cars can adapt to diverse road conditions, weather patterns, and traffic scenarios, learning to make split-second decisions that prioritize passenger safety.

In summary, reinforcement learning represents a groundbreaking approach to creating autonomous systems that can learn, adapt, and excel in complex environments. Whether it's a robot navigating challenging terrain, a drone performing acrobatic maneuvers, or a virtual assistant customizing content recommendations, reinforcement learning empowers machines to continuously improve their decision-making and performance. The future holds exciting possibilities as reinforcement learning continues to drive innovation across various domains, transforming the way we interact with intelligent machines. Machine learning has revolutionized the field of robot control, ushering in an era of intelligent, adaptable, and highly capable robotic systems. In the context of robot control, machine learning refers to the use of algorithms and models that allow robots to learn from data and improve their performance over time.

Imagine a scenario where a robot is tasked with picking and packing items in a warehouse. Traditionally, this task would require precise programming to handle various objects and their locations. However, with machine learning, the robot can continuously learn from its interactions, becoming more proficient at identifying, grasping, and placing items with each experience.

At the core of machine learning in robot control are two primary approaches: supervised learning and reinforcement learning. Supervised learning involves training a robot using labeled data, where each input is associated with a corresponding desired output. This method is particularly useful when the desired behavior is known and can be demonstrated.

Consider a scenario where a robot is learning to recognize and sort different types of fruit. In supervised learning, the robot is provided with a dataset of images of fruits, each labeled with the fruit's name. By analyzing these labeled images, the robot can learn to identify and categorize fruits accurately.

Reinforcement learning, on the other hand, takes a more exploratory approach. In this paradigm, robots learn through trial and error, receiving feedback in the form of rewards or penalties based on their actions. The robot's objective is to maximize its cumulative reward over time by discovering optimal behaviors.

Imagine a scenario where a robot is learning to navigate a maze. It starts by exploring various paths, receiving positive rewards for moving closer to the maze's exit and negative rewards for hitting walls or going in circles. Over time, the robot learns to select actions that lead it to the maze's exit efficiently.

Machine learning enables robots to handle complex, high-dimensional data. For example, in the field of computer vision, robots equipped with cameras can learn to interpret visual information. They can recognize objects, track their movements, and even understand human gestures and expressions.

Think of a robot with a built-in camera system that assists with elderly care. The robot can learn to recognize when an elderly person needs help, whether it's getting out of a chair or fetching a glass of water. By analyzing visual cues, the robot can respond appropriately to provide assistance.

One of the key advantages of machine learning in robot control is adaptability. Robots can adapt to changing environments, tasks, and user preferences. This adaptability is particularly valuable in applications where robots interact with humans, such as service robots or healthcare assistants.

Consider a service robot in a restaurant that learns to assist with serving dishes and clearing tables. As the restaurant's layout and menu change over time, the robot can adapt its behavior to efficiently navigate the dining area and interact with both customers and staff.

Machine learning also enables robots to generalize their knowledge. Instead of being programmed for specific tasks, robots can acquire a broader set of skills and apply them in novel situations. This generalization is akin to how humans learn to apply their knowledge and skills in various contexts.

Imagine a household robot that can learn to perform a wide range of chores, from cleaning floors to washing dishes. By training on different tasks and learning from user interactions, the robot can become a versatile and helpful assistant, adapting to the unique needs of each household.

Deep learning, a subset of machine learning, has been particularly influential in advancing robot control. Deep neural networks, inspired by the structure of the human brain, excel at processing and extracting valuable information from high-dimensional data, such as images, audio, and sensor readings.

Consider a scenario where a robot uses deep learning to improve its speech recognition capabilities. By processing audio data from microphones, the robot can learn to understand spoken commands and respond accurately. This technology is crucial for voice-controlled robots and virtual assistants.

Machine learning also plays a vital role in robot perception. Robots equipped with various sensors, such as cameras, lidar, and depth sensors, can use machine learning to interpret the sensory data and make sense of their surroundings. This perception is essential for tasks like autonomous navigation and object manipulation.

Think of a self-driving car that relies on machine learning to detect and classify objects on the road. Using data from onboard sensors, the car can recognize pedestrians, other vehicles, and traffic signs. This perception enables the car to make informed decisions and navigate safely.

In the realm of reinforcement learning, robots can learn complex control policies for tasks with high-dimensional action spaces. This capability is particularly valuable in applications like robotic manipulation, where precise control of robotic arms and grippers is essential.

Imagine a robot in a factory that learns to assemble intricate electronic components. Through reinforcement learning, the robot can discover optimal grasping strategies and manipulation techniques, enabling it to assemble products with precision and efficiency.

Another significant application of machine learning in robot control is in human-robot interaction. Robots can learn to understand and respond to human gestures, facial expressions, and natural language. This ability enhances communication and collaboration between humans and robots.

Consider a scenario where a robot assistant learns to assist with cooking in a kitchen. By observing and learning from a human chef's actions and verbal instructions, the robot can actively participate in the cooking process. It can chop vegetables, mix ingredients, and follow recipe instructions, making cooking a collaborative and enjoyable experience.

Machine learning in robot control is not without its challenges. Training machine learning models often requires large amounts of data, and fine-tuning models for specific tasks can be time-consuming. Additionally, ensuring the safety and ethical use of machine learning in robotics remains a priority.

In summary, machine learning has ushered in a new era of intelligent and adaptable robots. Whether it's learning to recognize objects, navigate complex environments, or interact with humans, machine learning empowers robots to continuously improve their capabilities. As technology advances, the role of machine learning in robot control will continue to expand, leading to more capable and versatile robotic systems that enhance our daily lives.

Chapter 7: Human-Robot Interaction and Collaboration

Designing natural and intuitive interfaces for robots is a critical aspect of ensuring that humans can interact with these machines effortlessly. It's akin to teaching robots to speak our language, but in this case, the language is a combination of physical movements, gestures, and sensory cues. The goal is to make human-robot interaction feel as natural as possible, removing barriers and making it accessible to individuals of all ages and backgrounds.

When we think about natural and intuitive interfaces for robots, we often envision a future where robots seamlessly integrate into our daily lives. Imagine a robot in your home that can understand your gestures, respond to your voice commands, and even anticipate your needs. It's a vision of technology that feels less like a machine and more like a helpful companion.

One of the fundamental elements of designing natural interfaces for robots is understanding human communication and behavior. Humans convey a vast amount of information through body language, facial expressions, and tone of voice. For robots to be effective communicators, they need to recognize and interpret these cues accurately.

Consider a robot designed to assist elderly individuals in their homes. To provide the best care and support, the robot must be able to perceive when someone is in distress or in need of help. This requires not only the ability to recognize physical signals, such as a person's posture or facial expression but also an understanding of context.

Context is a crucial factor in designing natural interfaces. People often communicate differently in various situations. For instance, a friendly chat with a robot at home might involve casual language and gestures, while a robot assisting in a professional setting should adopt a more formal tone.

Designing robots that can adapt their communication style to different contexts is essential for their integration into various aspects of our lives.

Voice recognition technology has made significant strides in recent years, enabling robots to understand and respond to spoken commands more accurately. This technology has made voice-controlled devices and virtual assistants like Siri and Alexa commonplace in our homes. For many, interacting with these devices has become second nature.

Imagine a scenario where a robot in your kitchen can understand your voice commands and assist with cooking. You can simply say, "Robot, preheat the oven to 350 degrees," and the robot responds by setting the oven temperature accordingly. Voice interfaces like these make it convenient for users to interact with robots in a hands-free and natural way.

Another critical aspect of natural interfaces is gesture recognition. Humans often use hand gestures to convey ideas, emotions, and instructions. Designing robots that can recognize and interpret these gestures is crucial for effective human-robot interaction.

For example, a robot in a factory might work alongside human employees. To coordinate their efforts efficiently, the robot should be able to understand and respond to hand signals given by the workers. This level of communication ensures a smooth and safe collaboration between humans and robots in industrial settings.

In the realm of healthcare, robots are being designed to assist patients with physical therapy exercises. These robots can recognize the movements of a patient's limbs and provide real-time feedback to ensure that exercises are performed correctly. By understanding and responding to the patient's movements, these robots create a more engaging and effective rehabilitation experience.

Facial expression recognition is another area of research and development in natural interfaces. Robots equipped with cameras and sophisticated algorithms can analyze a person's facial expressions to gauge their emotional state. This ability is particularly valuable in healthcare and mental health support applications.

Imagine a robot designed to provide companionship to individuals with autism. By analyzing the user's facial expressions and tone of voice, the robot can adapt its behavior to provide comfort and support. It can detect signs of distress and respond with calming gestures and words, creating a more nurturing and understanding interaction.

The use of touch and tactile feedback is another dimension of natural interfaces for robots. Some robots are equipped with sensors and actuators that allow them to sense and respond to touch. This capability enables robots to provide physical feedback and engage in gentle interactions.

For instance, consider a robot designed to assist in physical therapy for stroke patients. The robot can guide the patient through exercises and provide resistance when needed. It can sense the patient's movements and adjust its level of assistance accordingly, creating a personalized and responsive therapy experience.

Incorporating natural interfaces into robots also involves developing user-friendly control systems. Users should be able to interact with robots effortlessly, whether through a touchscreen interface, a mobile app, or a combination of voice and gesture commands. Designing intuitive control interfaces simplifies the user experience and encourages more widespread adoption of robotic technology.

Personalization is a key aspect of designing natural interfaces. Robots should be able to adapt to the unique preferences and needs of their users. This personalization can range from customizing the robot's name and appearance to tailoring its responses and behaviors.

Consider a robot designed for educational purposes. In a classroom setting, the robot can adapt its teaching style and content to the learning pace and preferences of individual students. It can provide personalized feedback and encouragement, making the learning experience more engaging and effective.

Ensuring the privacy and security of users is paramount when designing natural interfaces for robots. Robots equipped with cameras and microphones must handle data responsibly and protect users' information. Implementing robust security measures and providing transparency about data usage builds trust between users and robots.

In summary, designing natural and intuitive interfaces for robots is a multifaceted endeavor that draws from various fields, including artificial intelligence, computer vision, and human-computer interaction. The goal is to create robots that can understand and respond to human communication and behavior in a way that feels natural and seamless. As technology continues to advance, we can expect to see increasingly sophisticated and capable robots that enhance our lives through intuitive and empathetic interactions.

Ethical considerations in human-robot interaction are a crucial and evolving aspect of the rapidly advancing field of robotics. As we integrate robots into various aspects of our lives, it becomes imperative to address the ethical implications that arise from these interactions. This chapter will explore some of the key ethical considerations that researchers, engineers, and policymakers must take into account when developing and deploying robots in society.

One of the foremost ethical considerations in human-robot interaction is the question of robot autonomy and decision-making. As robots become more sophisticated and capable of autonomous actions, we must grapple with issues related to responsibility and accountability. For example, in situations where robots are entrusted with critical tasks, who bears the responsibility when something goes wrong? Is it the robot's designer, the operator, or both? Establishing clear lines of responsibility and accountability is essential to ensure ethical use of robotic technology.

Another ethical concern is the potential for robots to perpetuate or reinforce existing biases and inequalities. Machine learning algorithms, which underpin many aspects of robot behavior, can inherit biases present in the data they are trained on. This can lead to discriminatory or unfair outcomes. For instance, in the context of hiring or lending decisions, if a robot's decision-making process is biased against certain demographic groups, it can perpetuate discrimination. Addressing bias in robot decision-making is a critical ethical challenge that requires ongoing vigilance.

Privacy is another paramount ethical consideration in human-robot interaction. Robots equipped with sensors, cameras, and microphones can collect vast amounts of data about individuals and their environments. This data can include sensitive information about a person's daily routines, health status, or personal preferences. Protecting this data from unauthorized access or misuse is essential to safeguard individuals' privacy and maintain trust in robotic technology.

In healthcare settings, for example, robots may assist with patient care and data collection. It's imperative to establish robust privacy protocols and encryption mechanisms to protect patients' medical information and ensure compliance with healthcare privacy laws.

The issue of consent in human-robot interaction is closely tied to privacy. Individuals should have the right to consent to or decline robot interaction, particularly when it involves data collection or physical contact. Consent becomes especially relevant in situations where robots are used in sensitive or intimate contexts, such as caregiving or therapy.

Consider a scenario where a robot is designed to provide companionship and support to elderly individuals in their homes. While some may welcome the presence of such a robot, others may prefer not to interact with it. Respecting individuals' autonomy and choices by ensuring that they have the option to opt in or opt out of robot interactions is a key ethical consideration.

Transparency in robot behavior and decision-making is vital for building trust with users. Users should have a clear understanding of how robots operate and make decisions. If a robot is providing recommendations or assistance, users should be informed about the basis for those recommendations. Transparency builds confidence and helps users make informed decisions about how to interact with robots.

Additionally, transparency extends to the disclosure of the robot's capabilities. Users should be aware of the robot's sensory capabilities, data collection practices, and any limitations it may have. This transparency empowers users to make informed choices and fosters a sense of trust in the technology.

Ensuring that robots are designed with safety as a top priority is an ethical imperative. Robots operating in various environments, including homes, factories, and healthcare facilities, must be engineered to minimize the risk of harm to humans and other entities. This includes safety features such as collision detection and avoidance, emergency shutdown mechanisms, and fail-safe protocols.

In healthcare robotics, for example, robots that assist with surgeries must adhere to stringent safety standards to prevent harm to patients. Any robotic system that interacts with humans should undergo rigorous safety testing and certification processes to ensure it meets established safety standards.

The ethical dimension of job displacement is a topic of significant concern as automation and robotics continue to advance. While robots can improve efficiency and productivity, they can also lead to job displacement in certain industries. Addressing the potential economic and social impacts of automation is an ethical imperative. This may involve strategies such as job retraining programs, labor market adjustments, and policies to ensure a just transition for workers affected by automation.

Moreover, considerations of fairness come into play when robots are used in decision-making processes that impact individuals' lives. For instance, in legal contexts, algorithms and robots may be used to assist judges in making bail or sentencing decisions. Ensuring that these algorithms are fair and do not discriminate against particular groups is crucial for upholding principles of justice.

The ethical use of robots in military and defense applications is a contentious issue. Autonomous weapons, sometimes referred to as "killer robots," raise profound moral questions. The development and deployment of robots with the capacity to make lethal decisions without human intervention raise concerns about the potential for misuse and unintended consequences. International discussions and agreements are ongoing to establish ethical guidelines for the use of autonomous weapons.

In the realm of social robots designed for companionship and emotional support, ethical considerations revolve around the potential for robots to foster meaningful human relationships. While robots can simulate empathy and companionship to some extent, they do not possess genuine emotions or consciousness. It is essential to ensure that individuals do not replace human relationships with robotic ones and that robots complement human interactions rather than substitute for them.

As robots become more integrated into healthcare, there is also a need to consider the ethical implications of robot-assisted caregiving. While robots can provide valuable assistance and companionship to individuals with physical or cognitive impairments, they should not replace the essential human touch and empathy that caregivers provide. Striking the right balance between human and robotic care is an ethical challenge.

Finally, ethical considerations in human-robot interaction extend to questions of robot rights and personhood. As robots become increasingly advanced, the question arises: should robots have legal rights or some form of personhood status? This topic is a subject of ongoing debate and philosophical exploration, with no clear consensus.

In summary, ethical considerations in human-robot interaction are multifaceted and evolving. Addressing these ethical challenges requires interdisciplinary collaboration among roboticists, ethicists, policymakers, and society at large. As robots continue to play an increasingly prominent role in our lives, it is crucial to uphold ethical principles that prioritize human well-being, autonomy, and fairness while harnessing the potential benefits of robotic technology.

Chapter 8: Advanced Robot Sensing Technologies

Lidar and 3D sensing technologies are integral components of modern robotics, revolutionizing the way robots perceive and interact with their environments. These technologies, inspired by the human visual system, enable robots to sense and navigate complex spaces, avoid obstacles, and perform tasks with a level of precision and accuracy that was once considered the stuff of science fiction.

At the heart of 3D sensing in robotics lies Lidar, which stands for Light Detection and Ranging. Lidar systems work on a simple yet powerful principle: they emit laser pulses and measure the time it takes for those pulses to bounce back after hitting objects in the environment. By precisely calculating the time-of-flight of these laser pulses, Lidar sensors can create detailed and accurate 3D maps of the surroundings.

One of the key advantages of Lidar is its ability to provide high-resolution, real-time data about the robot's surroundings. This data includes information about the shapes, distances, and positions of objects, which is essential for navigation, obstacle avoidance, and manipulation tasks. Lidar's rapid data acquisition rate makes it well-suited for dynamic and fast-moving environments.

In addition to its real-time capabilities, Lidar excels in various environmental conditions. Unlike some other sensors, such as cameras, Lidar is not affected by changes in lighting conditions, making it highly reliable for outdoor applications where lighting can vary significantly. This robustness is particularly valuable in autonomous vehicles, where Lidar helps ensure safe navigation in diverse weather conditions.

The applications of Lidar in robotics are vast and continue to expand. In autonomous cars, Lidar sensors play a critical role in enabling vehicles to perceive and respond to their surroundings. These sensors provide the vehicle's AI system with a detailed 3D view of the road, allowing it to identify other vehicles, pedestrians, cyclists, and various obstacles. Lidar also helps the vehicle's navigation system create accurate maps and localize the vehicle within those maps.

Beyond autonomous vehicles, Lidar is used in a range of robotic systems, including drones, industrial robots, and service robots. In agriculture, Lidar-equipped drones can perform precision mapping and crop analysis, aiding farmers in optimizing their fields. In warehouses and manufacturing facilities, Lidar helps robots navigate cluttered spaces and safely collaborate with human workers.

However, while Lidar is a powerful technology, it is not without its challenges. One of the primary challenges is cost. High-quality Lidar sensors can be expensive, which can limit their widespread adoption, especially in consumer robotics. Researchers and engineers are actively working to develop more affordable Lidar solutions to address this barrier.

Another challenge is the need for robustness in various environmental conditions. While Lidar is generally unaffected by lighting conditions, it can face difficulties in adverse weather conditions like heavy rain or fog. Researchers are working on improving Lidar's performance under such conditions, as this is critical for applications like autonomous driving.

The size and form factor of Lidar sensors are also areas of ongoing innovation. As robotics applications become more diverse, there is a growing need for smaller, lighter, and more compact Lidar sensors that can be integrated into a wide range of robots, from drones to household robots. Miniaturization efforts are making progress in this direction.

In addition to Lidar, there are alternative 3D sensing technologies that complement and, in some cases, compete with Lidar. Stereo cameras, for example, use two cameras to capture depth information by triangulating the position of objects based on disparities in their images. While stereo cameras offer cost-effective 3D sensing, they may struggle in low-light or featureless environments.

Structured light is another 3D sensing technique that projects a known pattern onto objects and uses the deformation of that pattern to calculate depth. This technology is often used in industrial settings for tasks like quality control and object recognition.

Time-of-flight (ToF) cameras, also known as depth cameras, work by emitting light and measuring the time it takes for the light to return from objects in the scene. ToF cameras are found in consumer devices like gaming consoles and smartphones and are increasingly used in robotics for their compact form factor and affordability.

Each of these 3D sensing technologies has its strengths and limitations, and the choice of which to use depends on the specific requirements of the robotics application.

In summary, Lidar and 3D sensing technologies are at the forefront of robotics, enabling robots to perceive and interact with their environments in unprecedented ways. Lidar, in particular, has found applications in autonomous vehicles, drones, industrial automation, and beyond. As the field of robotics continues to advance, we can expect further innovations in 3D sensing technologies, leading to more capable and versatile robots that can safely and effectively navigate our world.

In the ever-evolving field of robotics, one area that has captured the imagination of researchers and engineers is bio-inspired sensing, a concept rooted in the idea of drawing inspiration from the natural world to enhance robotic perception and capabilities. The natural world is a treasure trove of ingenious solutions that have evolved over millions of years, and bio-inspired sensing aims to harness these solutions to create more capable and adaptive robots.

One of the most fascinating aspects of bio-inspired sensing is the emulation of the sensory systems found in various animals and organisms. Nature has produced a wide array of specialized sensors that enable creatures to navigate, interact with their environments, and even communicate with one another. By studying these sensors and the underlying biological mechanisms, researchers have been able to develop robotic systems with enhanced perception and the ability to operate in complex and dynamic environments.

Take, for example, the world of insects. Insects such as bees and ants are known for their remarkable navigation abilities, often performing intricate tasks like finding food sources and returning to their nests with pinpoint accuracy. Much of their navigation prowess can be attributed to their compound eyes, which are made up of thousands of individual lenses, each capturing a small portion of the environment. Inspired by this design, researchers have developed artificial compound eyes for robots. These eyes provide a panoramic view of the surroundings, enabling robots to detect motion, recognize objects, and navigate through cluttered environments with agility.

Beyond compound eyes, bio-inspired sensing has also led to the development of robotic whiskers, inspired by the tactile sensors found in animals like rats and seals. These whisker-like sensors are incredibly sensitive to touch and vibrations, allowing robots to explore dark or confined spaces, detect nearby objects, and even map their surroundings through touch alone. This type of sensory capability is invaluable in scenarios where vision or other traditional sensing methods may be limited.

Another area of bio-inspired sensing focuses on auditory perception, drawing inspiration from the hearing abilities of animals like bats. Bats use echolocation, emitting high-frequency sound waves and listening to the echoes to navigate and locate prey. Researchers have developed bio-inspired sonar systems for robots, enabling them to perform tasks like obstacle avoidance and mapping based on sound reflections. Such systems have applications in robotics for environments where visual sensing may be impaired or insufficient.

Biological olfaction, the sense of smell, has also inspired advancements in robotic sensing. Insects like moths and bees are known for their remarkable ability to detect and follow chemical gradients in the environment. Researchers have developed electronic noses and sensors that mimic these abilities, allowing robots to detect and identify odors in applications ranging from environmental monitoring to detecting gas leaks or chemical hazards.

Perhaps one of the most exciting frontiers in bio-inspired sensing is the field of soft robotics, which draws inspiration from the flexibility and adaptability of natural organisms. Soft robots are designed to mimic the movements and behaviors of creatures like octopuses and jellyfish. These robots often use soft, flexible materials and can perform tasks that are challenging for traditional rigid robots, such as navigating through tight spaces, interacting safely with humans, and conforming to irregular shapes in their environment.

Bio-inspired sensing is not limited to the emulation of individual sensory organs. Researchers are also exploring the integration of multiple sensors and sensory modalities to create robots with multimodal perception, similar to how humans combine information from vision, touch, and sound to understand their surroundings. This approach allows robots to make more informed decisions and adapt to a wider range of scenarios.

The practical applications of bio-inspired sensing are diverse and continue to expand. In search-and-rescue operations, for instance, robots equipped with bio-inspired sensors can navigate through disaster-stricken environments with greater precision, detecting signs of life or hazards that may be hidden from view. In agricultural robotics, bio-inspired sensors can help robots identify and manipulate crops, leading to more efficient and sustainable farming practices. Even in healthcare, bio-inspired sensors are being used to develop robots that can safely interact with and assist patients, such as in rehabilitation therapy.

However, while bio-inspired sensing holds immense promise, it also presents its share of challenges. One of the primary challenges is the need to translate complex biological systems into practical engineering solutions. The natural world is a product of evolution, and replicating its intricacies in artificial systems can be a daunting task. Researchers must grapple with questions of scalability, energy efficiency, and the robustness of bio-inspired sensors in real-world conditions.

Another challenge is the interdisciplinary nature of bio-inspired sensing, which requires collaboration between biologists, engineers, and computer scientists. Understanding the biological mechanisms underlying sensory systems is crucial for translating them into functional robotic sensors.

In summary, bio-inspired sensing represents a captivating and rapidly advancing field in robotics. By drawing inspiration from the natural world, researchers are pushing the boundaries of robotic perception and enabling robots to operate in diverse and challenging environments. The future holds the promise of robots with increasingly sophisticated sensory capabilities, opening up new possibilities for applications in fields as varied as exploration, healthcare, agriculture, and disaster response. As researchers continue to unravel the mysteries of the natural world, the potential for bio-inspired sensing to transform the field of robotics is nothing short of extraordinary.

Chapter 9: Robotic Ethics and Legal Considerations

As we delve deeper into the realms of robotics, we inevitably encounter a critical aspect that cannot be ignored—ethics. The development and deployment of robotic systems raise a multitude of ethical considerations that require careful thought and consideration. In this chapter, we will explore the various ethical frameworks that guide the design, operation, and use of robotic systems, with a focus on ensuring that these machines align with our moral values and societal norms.

Ethics in robotics is not a new concept but has gained increasing prominence in recent years as robots become more integrated into our daily lives. It's important to recognize that robots, in various forms, are being used in industries as diverse as healthcare, manufacturing, transportation, and even in our homes. As these machines become more autonomous and capable, ethical questions arise about how they should behave and interact with humans.

One of the foundational principles in the ethical development of robotic systems is safety. Ensuring the safety of both humans and robots is of paramount importance. Safety guidelines and standards are established to minimize the risks associated with robotic operations. This includes considerations for physical safety (avoiding collisions or harm to humans), cybersecurity (protecting against hacking or malicious use), and system reliability (reducing the likelihood of technical failures).

Beyond safety, a key ethical concern is transparency. It's important for users and society at large to understand how a robotic system works and makes decisions. This transparency is essential for trust and accountability. Users should have insight into how decisions are made, what data is collected, and how it is used. Transparency also extends to the disclosure of potential biases in algorithms and the ethical implications of the robot's actions.

The principle of autonomy is another critical ethical dimension. When robots are designed to make decisions on their own, it raises questions about the extent of their autonomy and the boundaries that should be in place. For example, in healthcare, robotic surgical systems can assist surgeons, but the ultimate decisions about patient care should remain in the hands of medical professionals.

Robotic systems also raise concerns related to privacy. In an era of data collection and analysis, robots can gather substantial amounts of information about individuals and their environments. This data can be valuable for improving system performance but must be handled with care to protect privacy rights. Ethical guidelines dictate how data should be collected, stored, and used, with a focus on obtaining informed consent from users.

Fairness and equity are ethical principles that come into play when designing and deploying robotic systems. Bias, whether intentional or unintentional, can lead to unfair outcomes. In applications like hiring or lending, automated decision-making systems must be designed to avoid discrimination based on factors such as race, gender, or socioeconomic status.

An essential ethical framework in robotics involves accountability and responsibility. When things go wrong, as they sometimes do, it's crucial to establish clear lines of accountability. Who is responsible when a robot makes an error or causes harm? This is a complex question that requires legal and ethical considerations, as well as the development of appropriate liability frameworks.

Considerations about the impact of robots on employment and the economy are also part of the ethical landscape. While robots can increase efficiency and productivity, they can also displace human workers in certain industries. Ethical discussions center around the responsibility of governments and organizations to retrain and support affected workers in transitioning to new roles.

The principle of benefit and harm is central to ethical decision-making in robotics. This involves evaluating the potential benefits of a robotic system against the potential harm it may cause. For example, autonomous vehicles have the potential to reduce accidents and save lives, but they also raise concerns about job displacement for drivers and the impact on the automotive industry.

Another ethical consideration is the cultural and societal context in which a robotic system operates. Norms and values can vary widely between different regions and communities. What is considered ethical behavior in one culture may be seen as unacceptable in another. This highlights the importance of adapting ethical guidelines to specific contexts and considering cultural diversity in the development and deployment of robotic systems.

The ethical frameworks we've discussed here are not exhaustive but provide a foundation for ethical decision-making in robotics. As the field continues to evolve, it's essential to engage in ongoing discussions and debates about the ethical implications of robotic technology. Ethical considerations should be an integral part of the development process, involving engineers, designers, ethicists, and policymakers.

In summary, ethics in robotics is a complex and multifaceted field that touches on safety, transparency, autonomy, privacy, fairness, accountability, economic impact, and cultural context. Ethical frameworks provide guidance for ensuring that robots align with our moral values and societal norms. By addressing these ethical considerations, we can harness the benefits of robotic technology while minimizing potential harm and fostering trust in these intelligent machines.

Privacy and security are paramount concerns in the ever-evolving landscape of robotics. As robots become more integrated into our daily lives, they interact with sensitive information and environments, making it essential to address the intricate web of privacy and security considerations that arise.

One of the primary privacy concerns in robotics revolves around data collection and usage. Robots, especially those equipped with sensors and cameras, can capture vast amounts of data about their surroundings and the people they interact with. This data may include images, audio recordings, and even personal information. The ethical use of this data is a critical aspect of privacy.

Ensuring that users have full control over their data is a fundamental principle. People should know what data is being collected by robots and how it will be used. This transparency is not only a legal requirement in many regions but also an ethical imperative. Users must be given the option to consent to data collection and, in some cases, have the ability to opt-out.

Data security is a closely related concern. When robots collect and store data, they become potential targets for malicious actors seeking to access or steal sensitive information. It is crucial to implement robust security measures to protect the data collected by robots. Encryption, secure storage, and access controls are among the strategies employed to safeguard this information.

In the context of healthcare, for example, robots that assist with patient care must handle medical data with the utmost care. Medical records are highly sensitive and subject to strict privacy regulations like the Health Insurance Portability and Accountability Act (HIPAA) in the United States. Robots involved in healthcare must comply with these regulations to protect patient privacy.

Another privacy consideration in robotics is the potential for surveillance and invasion of personal space. Robots equipped with cameras can inadvertently intrude on individuals' privacy, leading to discomfort and ethical concerns. This is particularly relevant in public spaces and workplaces, where robots may be used for security or monitoring purposes. Striking a balance between security and privacy is a delicate task.

Privacy by design is an approach that emphasizes incorporating privacy features into the development of robotic systems from the outset. This proactive approach seeks to minimize privacy risks by designating clear responsibilities for data handling and ensuring that privacy considerations are embedded in the system architecture.

Security concerns in robotics extend beyond data protection to encompass the physical security of the robot itself. Malicious actors may attempt to gain control of a robot to perform harmful actions. The consequences could range from property damage to endangering human safety. Implementing secure access controls and authentication mechanisms is essential to prevent unauthorized access.

Moreover, the integrity of robotic systems can be compromised by cyberattacks. Hackers may attempt to manipulate a robot's behavior, causing it to perform actions it wasn't designed for. In industrial settings, such attacks could lead to production disruptions, while in autonomous vehicles, they could result in accidents. Robust cybersecurity measures, including regular software updates and intrusion detection systems, are vital to mitigate these risks.

Ethical hacking or penetration testing is an approach used to assess the security of robotic systems. In this process, skilled professionals attempt to identify vulnerabilities in a robot's software and hardware. By uncovering weaknesses before malicious actors can exploit them, ethical hacking helps improve security.

Furthermore, the ethical use of robotics in security and law enforcement contexts is a contentious issue. The deployment of robots for surveillance, crowd control, or even lethal actions raises significant ethical questions. Policymakers and organizations must establish clear guidelines and oversight mechanisms to ensure that robots are used responsibly and in compliance with human rights.

Privacy and security concerns also intersect with the rise of social robots designed for companionship and assistance. These robots may interact with vulnerable populations, such as the elderly or children, raising concerns about data protection and emotional well-being. Striking a balance between providing companionship and ensuring privacy and emotional safety is an ongoing challenge.

In summary, privacy and security concerns are integral to the development and deployment of robotic systems. From data collection and usage to physical security and cybersecurity, the ethical use of robots hinges on addressing these complex considerations. Privacy by design, proactive cybersecurity measures, and ethical guidelines for deployment are essential components of responsible robotics. As technology continues to advance, it is crucial to prioritize privacy and security to build trust and ensure the positive integration of robots into society.

Chapter 10: Becoming a Robotics Specialist: Career and Research Paths

Selecting a specialization in the field of robotics is a significant decision that can shape your career and expertise. It's akin to choosing a path in a vast forest, with each direction leading to a unique destination. This chapter aims to help you navigate this decision-making process, exploring various robotics specializations, their significance, and the factors to consider when making your choice.

Robotics is an interdisciplinary field that encompasses a wide range of areas, each with its own set of challenges and opportunities. The first step in choosing your specialization is to gain a broad understanding of the key branches of robotics. This will enable you to make an informed choice that aligns with your interests and career goals.

One of the foundational areas in robotics is "Robot Design and Mechanics." In this specialization, you'll delve deep into the mechanical aspects of robots, learning how to design and build robotic systems from scratch. You'll explore materials, kinematics, dynamics, and other engineering principles that underpin the physical structure of robots. This specialization is ideal for those who enjoy hands-on work and have a passion for building tangible robotic devices.

Another exciting avenue is "Artificial Intelligence (AI) and Machine Learning for Robotics." This specialization focuses on imbuing robots with intelligence and autonomy. You'll delve into algorithms, deep learning, and reinforcement learning to teach robots how to perceive and interact with their environment. This area is at the forefront of robotics, driving advancements in autonomous vehicles, natural language processing, and computer vision.

"Robotics Perception and Sensing" is another specialization that plays a pivotal role in enabling robots to navigate and interact with the world. In this field, you'll explore the various sensors and perception systems that robots use to understand their surroundings. This includes computer vision, LIDAR, radar, and other technologies. Specializing in perception and sensing opens up opportunities in fields like autonomous driving, robotics in healthcare, and industrial automation.

For those interested in the intersection of robotics and healthcare, "Medical and Healthcare Robotics" is a compelling specialization. Here, you'll work on robots designed to assist medical professionals, enhance patient care, and perform complex surgeries. This specialization requires a solid understanding of both robotics and medical practices, making it a rewarding but challenging path.

"Industrial Robotics and Automation" is a specialization with immense practical significance. It focuses on designing and deploying robots for manufacturing and industrial applications. Industrial robots can perform tasks such as welding, assembly, and quality control with precision and efficiency. Specializing in this area can lead to opportunities in the manufacturing sector, where automation is rapidly transforming production processes.

"Human-Robot Interaction (HRI)" is a burgeoning field that explores how humans and robots can collaborate effectively and harmoniously. If you're interested in the social and psychological aspects of robotics, this specialization allows you to work on designing robots that can understand and respond to human emotions and behaviors. HRI has applications in fields like healthcare, education, and assistive technology.

If you have a penchant for exploring uncharted territories, "Space and Exploration Robotics" might pique your interest. This specialization involves developing robots for space exploration missions, such as those to Mars or the Moon. It's a field where robotics meets aerospace engineering, and it offers a chance to contribute to humanity's quest to explore the cosmos.

"Bio-Inspired Robotics" draws inspiration from nature to design robots with capabilities akin to living organisms. Specializing in this area allows you to explore how animals like birds, insects, and fish can inspire innovative robotic designs. This field has applications in areas like search and rescue, environmental monitoring, and swarm robotics.

Finally, "Ethical and Legal Considerations in Robotics" is an emerging specialization that deals with the ethical, legal, and societal implications of robotics. As robots become increasingly integrated into society, questions about responsibility, liability, and the impact on human lives arise. This specialization equips you to navigate the complex ethical and legal landscape of robotics.

When choosing your specialization, it's crucial to consider your interests, strengths, and long-term career goals. Take the time to explore each field, engage with professionals in those areas, and seek out educational opportunities. Additionally, consider the current trends and job prospects in your chosen specialization, as the field of robotics is continually evolving.

In summary, choosing your specialization in robotics is a pivotal step in your journey as a robotics enthusiast or professional. It's a decision that will shape your expertise and influence the contributions you make to the field. Whether you're drawn to the mechanical aspects of robot design, the intelligence of AI, the sensory capabilities of perception, or the ethical considerations surrounding robotics, there's a specialization that aligns with your passion and goals. Embrace the adventure of robotics, and remember that your chosen path is just the beginning of an exciting and ever-evolving journey.

Navigating the paths of academia and industry in the field of robotics can be an exhilarating yet challenging journey. These two distinct roads offer unique opportunities, experiences, and rewards, and the choice between them can significantly shape your career trajectory. In this chapter, we will explore the nuances of both academic and industry paths, helping you make an informed decision that aligns with your aspirations and ambitions.

Let's begin by delving into the academic path. Pursuing a career in academia means immersing yourself in the world of research, teaching, and intellectual exploration. As an academic, you will typically start by pursuing advanced degrees such as a Master's or Ph.D. in robotics or a related field. These degrees provide you with a deep understanding of the theoretical foundations and practical applications of robotics.

One of the most rewarding aspects of academia is the opportunity to contribute to the cutting edge of knowledge. As a researcher, you will engage in groundbreaking research projects, collaborate with experts in your field, and publish your findings in peer-reviewed journals and conferences. This process of discovery and innovation can be immensely fulfilling for those who have a passion for pushing the boundaries of what robots can do.

In addition to research, academia often involves teaching. You'll have the chance to educate the next generation of robotics enthusiasts and professionals, sharing your knowledge and insights with students eager to learn. Teaching can be a profoundly rewarding experience as you witness your students grow and develop their skills in robotics.

Furthermore, academia offers a level of autonomy and independence that some find appealing. As a professor or researcher, you have the freedom to choose your research topics, define your projects, and set your research agenda. This autonomy allows you to explore areas of robotics that align with your interests and expertise.

However, the academic path also comes with its challenges. Securing funding for research projects can be competitive, and there's a constant need to publish and maintain a strong research portfolio. Additionally, the tenure-track process, which leads to a permanent faculty position, can be rigorous and demanding.

Now, let's turn our attention to the industry path. Working in the robotics industry offers a dynamic and fast-paced environment where you can apply your skills to real-world problems. Many industries, from automotive manufacturing to healthcare, are increasingly relying on robotics to enhance efficiency and productivity.

One of the primary advantages of the industry path is the opportunity for practical, hands-on work. In industry roles, you'll design, develop, and deploy robotic systems that have a direct impact on the products and services offered by your organization. Whether you're building autonomous vehicles, medical robots, or warehouse automation systems, you'll see your work in action.

Industry positions often come with competitive salaries and benefits, making them financially attractive. Moreover, the demand for robotics professionals in industry is on the rise, offering a wide range of job opportunities and career advancement prospects.

Collaboration is a hallmark of the industry path. You'll work closely with multidisciplinary teams, including engineers, designers, and project managers, to bring robotic solutions to life. This collaborative environment fosters innovation and allows you to apply your expertise in a real-world context.

However, industry roles may also come with certain pressures and constraints. Project timelines, budgets, and customer demands can be challenging to manage. Moreover, the specific focus of your work may be determined by your employer's needs, limiting the scope of your research interests.

When contemplating your career path in robotics, it's essential to consider your personal goals and values. Reflect on what motivates you and where you see yourself making the most significant impact. Some individuals are drawn to the academic path, relishing the opportunity to contribute to the theoretical foundations of robotics. Others thrive in the fast-paced, applied environment of industry, where they can witness their innovations in action.

It's worth noting that the boundary between academia and industry is not rigid. Many robotics professionals find ways to bridge these two worlds, engaging in collaborative research projects with industry partners or pursuing industrial careers after academic stints. Such hybrid paths can offer the best of both worlds, combining the intellectual stimulation of academia with the practicality of industry.

Regardless of your chosen path, continuous learning and professional development are essential. Robotics is a rapidly evolving field, and staying up-to-date with the latest advancements and technologies is crucial. Seek out networking opportunities, attend conferences, and engage with the vibrant robotics community to nurture your growth.

In summary, the decision to pursue an academic or industry career in robotics is a significant one, and it should align with your interests, values, and career aspirations. Both paths offer unique rewards and challenges, and there is no one-size-fits-all answer. Whether you choose to embark on the academic journey of research and teaching or dive into the dynamic world of robotics industry, your contribution to the field is valuable, and your passion for robotics will continue to drive innovation and progress.

BOOK 4
MASTERING ROBOTICS RESEARCH
FROM ENTHUSIAST TO EXPERT

ROB BOTWRIGHT

Chapter 1: The Journey of a Robotics Enthusiast

Exploring your passion for robotics is a thrilling and intellectually stimulating journey that opens up a world of possibilities. Robotics is a multifaceted field that encompasses a wide range of disciplines, making it both challenging and rewarding to explore. In this chapter, we will delve into the exciting realm of robotics and guide you on a path to discovering and nurturing your passion for this transformative field.

At its core, robotics is the intersection of various scientific and engineering disciplines. It combines mathematics, physics, computer science, mechanical engineering, electrical engineering, and more to create intelligent machines capable of performing tasks autonomously or semi-autonomously. This multidisciplinary nature of robotics is what makes it so captivating.

The first step in exploring your passion for robotics is to gain a foundational understanding of its key components and principles. You'll want to familiarize yourself with the fundamental concepts that underpin robotic systems. Mathematics plays a critical role in robotics, from linear algebra for representing transformations to calculus for motion planning and control.

Physics principles are equally important, as they provide the foundation for understanding how robots interact with the physical world. Concepts such as kinematics, dynamics, and forces are essential for designing robots that can move, manipulate objects, and navigate their environments effectively.

As you delve deeper into the field, you'll encounter the various components that make up a robotic system. Robotic actuators, which are responsible for generating motion, come in various forms, including electric motors, pneumatic systems, and hydraulic actuators. Understanding the strengths and limitations of different actuators will help you make informed design choices.

Sensor technology is another critical aspect of robotics. Sensors provide robots with the ability to perceive and interact with their surroundings. From cameras for visual perception to lidar for 3D mapping, sensors enable robots to gather data about the world and make informed decisions based on that information.

Control systems are the brains behind robotic operations. They dictate how a robot's actuators should move in response to sensor data, allowing the robot to perform tasks accurately and efficiently. Learning about control theory and algorithms will be invaluable in your exploration of robotics.

Robotics also encompasses the field of artificial intelligence (AI) and machine learning. These technologies enable robots to adapt and learn from their experiences, making them more versatile and capable. Exploring AI and machine learning in the context of robotics will open doors to advanced applications such as autonomous navigation and decision-making.

With a solid foundation in the core principles and components of robotics, you can start experimenting with hands-on projects. Building your first robot, whether it's a simple wheeled robot or a more complex robotic arm, is a fantastic way to apply your knowledge and gain practical experience. There are numerous robotics kits and platforms available that cater to beginners and enthusiasts, making it accessible to anyone with an interest in robotics.

As you embark on your robotics journey, consider the specific areas of robotics that pique your interest. Do you have a fascination with autonomous drones and their applications in aerial photography or environmental monitoring? Are you drawn to humanoid robots that mimic human movement and interaction? Perhaps you're intrigued by the potential of robots in healthcare, from surgical robots to robotic prosthetics.

Exploring your passion for robotics also means engaging with the vibrant robotics community. Attend robotics conferences, join online forums and communities, and connect with like-minded individuals who share your enthusiasm. Collaborating with others can lead to exciting projects, knowledge sharing, and a deeper appreciation for the field.

Another crucial aspect of exploring your passion for robotics is staying informed about the latest developments and trends. Robotics is a rapidly evolving field, with breakthroughs happening regularly. Reading research papers, following robotics news, and subscribing to robotics journals will help you keep up-to-date with the cutting-edge advancements in the field. Furthermore, consider pursuing formal education in robotics if you're looking to deepen your knowledge and skills. Many universities offer undergraduate and graduate programs in robotics or related fields. These programs provide structured learning experiences, access to cutting-edge research, and opportunities to collaborate with leading experts in the field.

Internships and research opportunities with robotics companies or research institutions can also provide valuable hands-on experience and insights into the practical applications of robotics. Working on real-world projects will give you a taste of what it's like to be part of the robotics industry.

In your exploration of robotics, don't be afraid to embrace challenges and setbacks. Building and programming robots can be challenging, and you'll likely encounter obstacles along the way. However, overcoming these challenges is part of the learning process and can be incredibly rewarding.

Lastly, remember that your passion for robotics can have a positive impact beyond your personal interests. Robotics has the potential to address some of society's most pressing challenges, from assisting in disaster response to revolutionizing healthcare and transportation. By pursuing your passion for robotics, you may contribute to advancements that benefit humanity as a whole.

Embarking on a journey from a robotics hobbyist to an aspiring researcher is a path filled with exciting challenges and opportunities. Whether you've been building robots as a hobby for years or have just discovered your passion for robotics, this chapter will guide you on the transformative journey toward becoming a researcher in the field.

Your journey begins with a strong foundation in the fundamental concepts of robotics. While you may have gained some knowledge through your hobbyist projects, it's essential to delve deeper into the theoretical aspects of robotics. This includes understanding the mathematical principles that underlie robotic systems.

Linear algebra, for example, plays a crucial role in robotics, as it provides the mathematical framework for representing transformations and coordinate transformations in three-dimensional space. A solid grasp of linear algebra will allow you to work with robot kinematics, dynamics, and control more effectively. Calculus and differential equations are equally important, especially when it comes to motion planning and control. These mathematical tools enable you to describe and analyze the motion of robots in various environments. Whether you're designing a robot arm or programming a drone for autonomous flight, calculus and differential equations will be your allies. Probability and statistics are another pair of indispensable tools in a researcher's toolkit. They are essential for modeling uncertainty, which is inherent in many robotic systems. Bayesian probability, in particular, is widely used in robotics for tasks such as localization and sensor fusion.

Once you've established a solid mathematical foundation, it's time to dive into the core areas of robotics research. Kinematics and dynamics of robotic systems are fundamental topics that researchers explore in depth. Kinematics deals with the study of motion without considering the forces involved, while dynamics incorporates the forces and torques that affect a robot's motion. These areas are essential for understanding how robots move and interact with their environment.

Programming and software development are critical skills for a robotics researcher. While you may have programmed robots for your hobby projects, research-level programming often involves more complex algorithms and the use of specialized robotic software frameworks. Learning to work with ROS (Robot Operating System) or other similar platforms will be valuable on your research journey.

Machine learning and artificial intelligence (AI) have become integral to robotics research. These technologies enable robots to learn from data, adapt to changing environments, and make intelligent decisions. Exploring advanced machine learning techniques and AI algorithms will open up exciting possibilities for your research projects.

Sensor fusion is another area of interest for robotics researchers. Combining data from multiple sensors, such as cameras, lidar, and inertial sensors, allows robots to create more accurate and comprehensive representations of their surroundings. Understanding sensor fusion techniques is essential for enhancing a robot's perception capabilities.

As you transition from a hobbyist to a researcher, you'll likely find that your projects become more ambitious and research-oriented. Research projects often involve designing and building robots from scratch or customizing existing platforms to suit specific experimental needs. Collaborating with mentors, advisors, or research teams can provide valuable guidance and resources for your projects.

Conducting experiments and collecting data is a crucial aspect of robotics research. You'll need to design experiments that test your hypotheses and gather data to analyze and draw conclusions. Tools such as simulation software, data acquisition systems, and laboratory equipment will aid in your research endeavors.

Publishing your research findings in academic journals and presenting them at conferences is an essential step in becoming a recognized researcher in the field. It allows you to share your insights, contribute to the robotics community, and receive feedback from peers. It's also an opportunity to establish your reputation as a researcher.

Networking within the robotics community is invaluable for your research journey. Attend conferences, workshops, and seminars to connect with fellow researchers, professors, and industry professionals. Building a network of collaborators and mentors can provide you with guidance, access to resources, and potential research opportunities. Consider pursuing advanced degrees, such as a master's or a Ph.D., to formalize your research training. Graduate programs offer the opportunity to work on cutting-edge research projects, collaborate with faculty members, and access research facilities and funding. As you progress in your journey from hobbyist to researcher, you may discover specific areas of robotics that resonate with your interests. Whether it's computer vision, human-robot interaction, swarm robotics, or any other subfield, focusing your research on a niche area can lead to in-depth expertise and groundbreaking contributions.

Throughout your journey, perseverance and resilience are key. Research can be challenging, and setbacks are a natural part of the process. However, each obstacle you overcome brings you closer to achieving your research goals.

In summary, the transition from a robotics hobbyist to an aspiring researcher is an exciting and fulfilling journey. It involves deepening your knowledge of mathematical and theoretical foundations, exploring core areas of robotics research, honing your programming and software development skills, and engaging in hands-on research projects. With dedication, passion, and continuous learning, you can make significant contributions to the ever-evolving field of robotics.

Chapter 2: Refining Your Robotics Research Interests

Identifying your niche in the vast field of robotics is a pivotal step in your journey as a robotics enthusiast and researcher. While robotics as a whole encompasses a wide range of technologies and applications, narrowing down your focus to a specific niche can lead to a more rewarding and impactful experience. In this chapter, we'll explore the process of discovering and honing your niche in robotics.

The world of robotics is brimming with exciting possibilities, and it's common for individuals to have diverse interests within the field. Your first task is to reflect on your passions and curiosities. What aspects of robotics spark your enthusiasm? Is it the development of autonomous vehicles, the creation of humanoid robots, or perhaps the application of robotics in healthcare?

Consider your background and expertise. Are there any subjects or fields you are particularly knowledgeable in or passionate about? For example, if you have a strong background in computer science, you might be drawn to robotics applications that heavily rely on software and algorithms, such as robot perception or machine learning for robotics.

Your career aspirations can also guide you toward a specific niche. If you dream of contributing to the future of space exploration, you might find your niche in space robotics, which involves designing and building robots for extraterrestrial exploration and colonization.

Another avenue to explore is your personal experiences and challenges. Have you encountered a problem in your daily life or witnessed a societal issue that could be addressed through robotics? These real-world challenges can lead to niche areas where your expertise can make a significant impact.

It's essential to stay informed about the latest trends and developments in robotics. Attend conferences, workshops, and seminars in your areas of interest to connect with experts and learn about emerging technologies and research directions. Engaging with the robotics community can help you discover niche areas that align with your passion and skills.

As you explore potential niches, keep in mind that interdisciplinary approaches are often the most innovative. Robotics is inherently multidisciplinary, drawing from fields like computer science, electrical engineering, mechanical engineering, and more. Don't hesitate to combine your expertise with knowledge from other domains to create unique solutions.

Collaboration is a powerful tool for niche discovery. Joining robotics research teams or working on collaborative projects can expose you to a wide range of subfields. Interacting with teammates and mentors can help you identify areas where you excel and find your niche.

Consider the societal impact of your chosen niche. How does it align with your values and goals? Robotics has the potential to address critical global challenges, such as environmental sustainability, healthcare accessibility, and disaster response. Choosing a niche that resonates with your values can be a source of motivation and fulfillment.

As you narrow down your options, conduct a thorough review of the existing literature and research in your chosen niche. This will help you understand the current state of the field, identify gaps in knowledge, and pinpoint areas where you can make meaningful contributions.

Don't shy away from exploring unconventional or emerging niches. The field of robotics is constantly evolving, and new opportunities arise as technology advances. Consider areas like soft robotics, biohybrid systems, or swarm robotics, which are gaining traction as exciting frontiers in the field.

Once you've identified your niche, it's time to dive deep into the subject matter. Immerse yourself in research, engage in hands-on projects, and seek mentorship from experts in your chosen area. Building expertise and a strong foundation is essential to becoming a respected contributor in your niche.

Networking within your niche is crucial. Attend specialized conferences and workshops dedicated to your area of interest. Connect with researchers, practitioners, and enthusiasts who share your passion. Collaboration often leads to groundbreaking discoveries and projects that can shape the future of your niche.

Consider pursuing advanced degrees or certifications related to your niche. Formal education can provide you with the knowledge, skills, and credentials needed to excel in your chosen area. Additionally, it can open doors to research opportunities and collaborations with academic institutions.

Stay adaptable and open to evolving within your niche. As robotics technology advances, new subfields and applications may emerge. Being flexible and willing to explore new directions can keep your career in robotics dynamic and fulfilling.

In summary, identifying your niche in robotics is a journey of self-discovery and exploration. It involves reflecting on your passions, leveraging your expertise, and staying informed about emerging trends. Collaborating with mentors and peers, conducting thorough research, and networking within your niche are key steps toward becoming a proficient and influential contributor in your chosen area of robotics. Embrace the challenges and opportunities that your niche presents, and let your passion drive your journey forward in the exciting world of robotics.

Selecting your research focus is a pivotal step in your journey as a robotics specialist. It's a decision that will guide your work, shape your expertise, and contribute to the ever-evolving landscape of robotics. In this chapter, we'll explore the intricacies of selecting the right research focus, considering the multitude of options and the profound impact your choice can have.

Research in robotics is a dynamic and rapidly expanding field, encompassing a wide array of domains and subdomains. Before diving into the specifics, it's essential to recognize your broader interests within robotics. Are you drawn to the realm of autonomous vehicles, eager to develop innovative healthcare robots, or captivated by the potential of human-robot collaboration? Your passions and curiosities are the compass that will guide your research journey.

Once you've identified your overarching interests, it's time to delve deeper. Within each broad domain, there are numerous research topics and areas to explore. For example, if you're intrigued by autonomous vehicles, you might choose to specialize in self-driving cars, unmanned aerial vehicles (UAVs), or even autonomous underwater vehicles (AUVs). Each of these subfields offers its own set of challenges and opportunities.

To make an informed decision, immerse yourself in the existing body of research. Read scientific papers, attend conferences, and engage with the robotics community. This will give you a sense of the current state of research in various areas and help you identify gaps or unexplored avenues that align with your interests.

Consider the societal impact of your research focus. How does it contribute to solving real-world problems or advancing human capabilities? Robotics has the potential to address significant global challenges, from improving healthcare to mitigating the effects of climate change. Choosing a research focus that resonates with your values and has a positive societal impact can be a source of motivation and purpose.

Interdisciplinary approaches often lead to groundbreaking research. Robotics inherently draws from various fields, including computer science, mechanical engineering, electrical engineering, and more. Don't hesitate to blend your expertise with insights from other disciplines, as this can lead to innovative solutions and novel perspectives.

Collaboration is a powerful driver of research. Joining or forming research teams can expose you to diverse viewpoints and skill sets. Collaborative projects often tackle complex challenges that require expertise from multiple domains. Interacting with fellow researchers can help you refine your focus and generate fresh ideas.

Consider the long-term trajectory of your research focus. Is it an area that will remain relevant and promising in the coming years? Technological advancements, industry trends, and societal needs can influence the trajectory of research fields. Staying attuned to these factors can help you make informed decisions about your research focus.

Seek mentorship from experienced researchers in your chosen area. Mentors can provide valuable guidance, share their insights, and help you navigate the complexities of research. Their expertise and mentorship can accelerate your growth as a researcher and guide your focus.

Reflect on your strengths and skills. What unique contributions can you bring to your chosen research area? Identifying your strengths can help you carve out a niche within your focus and make meaningful contributions.

As you narrow down your research focus, remember that it's not a static decision. Research is an iterative process, and your focus may evolve over time as you gain new insights and experiences. Be open to adjusting your focus as needed to stay aligned with your goals and interests.

Once you've selected your research focus, it's time to define your research questions and objectives. What specific problems do you aim to address? What are the key research questions you want to answer? Developing a clear research plan is essential for making progress and communicating your goals to others.

Incorporate flexibility into your research journey. Unexpected discoveries and breakthroughs can lead you in new and exciting directions. Embrace the serendipitous aspects of research and be open to exploring uncharted territory.

In summary, selecting your research focus in robotics is a multifaceted process that involves aligning your passions, expertise, and societal impact considerations. It requires immersion in the field, collaboration, mentorship, and ongoing reflection. Your research focus will shape your contributions to the world of robotics, and by making thoughtful and informed choices, you can embark on a fulfilling and impactful research journey.

Chapter 3: Advanced Mathematical Concepts for Robotics

Linear algebra serves as the mathematical foundation that underpins much of robotics research. It's the mathematical language that robots and researchers use to communicate and model the physical world around them. At its core, linear algebra deals with vectors and matrices, and understanding its principles is crucial for various aspects of robotics, from kinematics and dynamics to computer vision and control.

Vectors, often represented as arrows in three-dimensional space, are fundamental in robotics. They describe quantities that have both magnitude and direction. In robotics, vectors can represent various physical entities, such as the position and orientation of a robot's end effector, the force applied to a robot's joint, or the velocity of a moving object. These vectors play a pivotal role in describing the state of a robot and its interaction with the environment.

Matrices are arrays of numbers arranged in rows and columns. They are used to transform vectors and perform operations in robotics. For instance, a transformation matrix can describe how a robot's end effector moves from one coordinate system to another, enabling it to navigate and manipulate objects in its environment. Additionally, matrices are used to represent the relationships between robot joints and their motion.

Matrix operations, such as multiplication, addition, and inversion, are fundamental in solving robotics problems. They allow researchers to model complex robot systems and manipulate data efficiently. For example, when calculating the forward kinematics of a robotic arm, matrix transformations are used to determine the position and orientation of the end effector based on joint angles.

Eigenvalues and eigenvectors are critical concepts in linear algebra for robotics. They help researchers understand the inherent properties of linear transformations, such as rotations and scaling. In robotics, eigenvalues and eigenvectors are used in various applications, including stability analysis of control systems and characterizing the behavior of robotic sensors.

Singular value decomposition (SVD) is another powerful tool in linear algebra with applications in robotics. SVD is used for tasks like data compression, noise reduction, and solving systems of linear equations. In robotics, SVD plays a role in sensor calibration, robot motion planning, and optimization problems.

Robotics researchers use linear algebra to describe the motion of robotic systems accurately. The motion of robots, whether in manufacturing, healthcare, or exploration, is governed by mathematical equations that involve vectors and matrices. These equations allow researchers to predict how a robot will move and interact with its surroundings, which is essential for tasks like path planning, obstacle avoidance, and trajectory generation.

Robot kinematics, which deals with the study of robot motion without considering forces and torques, heavily relies on linear algebra. Researchers use transformation matrices to represent the configuration of robot joints and derive the end effector's position and orientation. This information is crucial for tasks like pick-and-place operations, where precise positioning is required.

Beyond kinematics, linear algebra also plays a central role in robot dynamics. Robot dynamics involves understanding the forces and torques that act on a robot as it moves. Researchers use matrices to model the relationships between joint forces, joint accelerations, and robot motion. This knowledge is essential for designing control algorithms that ensure the robot moves safely and efficiently.

Computer vision is a field of robotics that focuses on enabling robots to interpret and understand visual information from the world around them. Linear algebra is a key component in image processing and computer vision algorithms. Matrices are used to perform operations like image filtering, feature extraction, and image transformations. These operations are fundamental for tasks such as object recognition, tracking, and navigation.

Control theory, which deals with regulating the behavior of robotic systems, relies heavily on linear algebra. Researchers use matrices to represent the dynamics of a robot and design control algorithms that steer the robot towards desired goals while accounting for external disturbances. Linear algebra provides the mathematical tools needed to analyze the stability and performance of control systems.

In summary, linear algebra is a cornerstone of robotics research, providing the mathematical framework for modeling, analyzing, and controlling robotic systems. From representing robot configurations and transformations to enabling computer vision and control, linear algebra empowers researchers to tackle the diverse challenges of robotics, making it an indispensable tool in the field.

Calculus and differential equations are the mathematical workhorses that power many aspects of robotics research and development. These mathematical tools are essential for understanding and describing how robotic systems move, sense, and interact with their environment. While they might seem daunting to some, they are the building blocks of robotics, enabling us to navigate complex challenges in the field.

Calculus, a branch of mathematics developed by Isaac Newton and Gottfried Wilhelm Leibniz in the 17th century, is divided into two main branches: differential calculus and integral calculus. Differential calculus focuses on understanding rates of change and slopes of curves, while integral calculus deals with the accumulation of quantities and areas under curves. Both branches are indispensable in robotics.

Differential calculus, in particular, plays a vital role in modeling the behavior of robotic systems. It allows us to describe how quantities change concerning one another. For example, when considering the motion of a robot, we use calculus to calculate velocities and accelerations at different points along its path. These calculations are fundamental for tasks like trajectory planning and control.

Velocity, the rate of change of position, is a key concept in robotics, and calculus provides us with the tools to analyze and control it. Whether a robot is moving along a predetermined path or navigating through an unknown environment, understanding how velocity changes over time is crucial for ensuring safe and efficient movement.

Acceleration, the rate of change of velocity, is equally important. Robotics researchers use calculus to compute accelerations in various contexts, such as when designing robots that need to accelerate or decelerate smoothly to avoid sudden jerks or instability.

Integral calculus, on the other hand, comes into play when we need to compute quantities that accumulate over time or space. In robotics, this is often related to sensor data. For example, when a robot uses a distance sensor to measure the distance it has traveled, integral calculus helps us calculate the total distance covered by integrating the sensor's measurements over time.

Differential equations are another mathematical tool that is indispensable in robotics. These equations describe how a quantity changes concerning one or more other quantities. In robotics, they are used to model the dynamics of robotic systems. For instance, when a robot manipulator moves its joints, differential equations help us understand how the joint angles change concerning time and how these changes relate to the motion of the end effector.

These equations are particularly relevant in control theory, where they are used to design control algorithms that regulate the behavior of robotic systems. Engineers and researchers employ differential equations to formulate and solve problems related to robot control, ensuring that robots can follow desired trajectories or achieve specific tasks accurately.

Moreover, in the context of robotics perception, differential equations play a role in sensor fusion. When a robot integrates data from multiple sensors to create a comprehensive understanding of its environment, differential equations can help in synchronizing and aligning the data streams, enabling the robot to build an accurate representation of the world around it.

In computer vision, which is vital for many robotics applications, differential equations come into play when dealing with image processing tasks like edge detection or motion estimation. These equations help researchers create algorithms that allow robots to interpret visual information, track objects, and navigate autonomously.

The fundamental principles of calculus and differential equations provide the groundwork for solving complex problems in robotics. Whether it's designing robots that can move gracefully and precisely, developing control algorithms that keep robots stable and responsive, or enabling robots to perceive and understand their surroundings, calculus and differential equations are the mathematical tools that empower us to overcome the challenges of robotics.

So, while these mathematical concepts may appear abstract or intimidating, they are the very tools that allow us to turn our robotic dreams into reality. They enable us to create machines that can assist us in various aspects of our lives, from manufacturing and healthcare to exploration and entertainment. Robotics is a multidisciplinary field, and calculus and differential equations are the common languages that bridge the gap between theory and practical applications, making robots not just a product of science fiction but an integral part of our present and future.

Chapter 4: Robotic Sensing and Perception at an Expert Level

Perception algorithms lie at the heart of enabling robots to interact effectively with complex and dynamic environments. In the world of robotics, perception refers to the ability of a robot to sense and interpret information from its surroundings, much like our human senses of sight, touch, and hearing. These algorithms allow robots to gather data from various sensors and make sense of it, which is essential for tasks ranging from navigation to object manipulation.

One of the fundamental challenges in developing perception algorithms for robots is dealing with the inherent uncertainty in sensory data. In the real world, sensors can be noisy, and the environment can change unpredictably. Perception algorithms need to account for these uncertainties and make sense of imperfect data. Think of it as trying to read a smudged message or interpret a blurry image; the algorithms need to decipher the meaningful information from the noise.

Computer vision is a significant component of perception in robotics. It involves the use of cameras and image processing techniques to help robots "see" and understand their environment. For example, robots equipped with cameras can capture images or video streams, and perception algorithms then analyze these visual data to identify objects, track their movements, and even understand their spatial relationships.

In recent years, deep learning techniques, particularly convolutional neural networks (CNNs), have revolutionized computer vision in robotics. These networks have demonstrated remarkable capabilities in image classification, object detection, and image segmentation, making it possible for robots to perform more advanced tasks. For instance, a robot with a CNN-based perception system can recognize a wide range of objects, from everyday items to specific tools required for particular tasks.

But perception in robotics goes beyond just vision; it encompasses other senses as well. For example, robots can be equipped with various sensors such as lidar, radar, and ultrasonic sensors to perceive their surroundings in three dimensions. Lidar, which uses laser beams to measure distances to objects, is particularly crucial for creating detailed maps of the environment. Robots in fields like autonomous vehicles rely heavily on lidar-based perception for navigation and obstacle avoidance. Another essential aspect of perception is sensor fusion. This involves combining data from multiple sensors to create a more comprehensive and accurate understanding of the environment. For instance, a robot might fuse data from cameras, lidar, and IMUs (inertial measurement units) to navigate through complex environments. Sensor fusion algorithms must align and synchronize data from different sensors to ensure that the robot's perception is consistent and coherent. Moreover, perception algorithms play a vital role in the localization of robots within their environment. Simultaneous localization and mapping (SLAM) is a classic problem in robotics where a robot must build a map of its surroundings while simultaneously determining its own position within that map. Perception algorithms help robots achieve this by processing sensor data and identifying landmarks or features in the environment to update their position and map. Natural language processing is yet another area where perception algorithms are making significant strides in robotics. With the ability to understand and process spoken or written language, robots can interact more intuitively with humans. This is particularly valuable in scenarios where human-robot collaboration is essential, such as healthcare or customer service. One fascinating direction in perception research is bio-inspired sensing. Researchers draw inspiration from nature to develop sensors and algorithms that mimic biological systems. For example, some robots are equipped with tactile sensors that emulate human touch, allowing them to perform delicate tasks or respond to physical interactions in a more human-like manner. The challenges in perception algorithms are not only technical but also ethical and societal. As robots become more integrated into our lives, ensuring that they perceive and interact with their environment ethically and safely is crucial. Privacy concerns, for instance, arise when robots are equipped with sensors that can capture personal data or images without consent. Developing algorithms that respect privacy while still enabling effective perception is a delicate balance.

In summary, perception algorithms are the cornerstone of enabling robots to navigate and interact with complex and dynamic environments. These algorithms leverage data from various sensors, including cameras, lidar, and IMUs, to provide robots with a sense of their surroundings. They play a crucial role in tasks ranging from autonomous navigation to object manipulation and natural language understanding. As robotics continues to advance, so too will the capabilities of perception algorithms, bringing us closer to a world where robots can seamlessly and intelligently coexist with humans. Robotic vision and object recognition are at the forefront of robotics research, representing a significant leap in a robot's ability to understand and interact with its environment. Imagine a robot that can see and understand the world around it, recognizing objects, identifying them, and even predicting their movements. This level of mastery in robotic vision and object recognition is a transformative capability, one that brings robots closer to being versatile and valuable companions in various applications.

Robotic vision is all about giving robots the sense of sight. Just like our eyes and visual processing system, robots equipped with cameras and advanced vision algorithms can capture and interpret visual data from their surroundings. These cameras act as the robot's eyes, allowing it to perceive the world in two or three dimensions.

Object recognition is a fundamental aspect of robotic vision, and it involves the ability to identify and categorize objects within a robot's field of view. Think about how we effortlessly recognize everyday objects—a coffee mug, a book, or a laptop. For a robot to achieve this level of understanding, it must process visual data, extract relevant features, and match them to known objects or categories.

At the heart of object recognition in robotics are deep learning techniques, particularly convolutional neural networks (CNNs). These neural networks can be trained on massive datasets containing images of various objects, enabling the robot to learn to recognize thousands of objects with high accuracy. The process involves training the network to detect specific features or patterns within the images, such as the shape, color, or texture of objects.

Once a robot has been trained in object recognition, it can identify objects in real-time, even in complex and cluttered environments. For example, a robot in a warehouse can identify products on shelves, a home robot can recognize and locate items in a cluttered room, or an autonomous vehicle can detect pedestrians and other vehicles on the road.

Object recognition extends beyond simple identification; it also includes understanding the spatial relationships between objects. This ability, known as scene understanding, allows a robot to grasp not only what objects are present but also how they relate to one another. For instance, it can discern that a cup is on top of a saucer, or that a book is on a shelf.

One of the exciting advancements in robotic vision is the integration of 3D perception. Instead of perceiving the world in flat images, robots can now create three-dimensional representations of their surroundings. This is often achieved using depth-sensing cameras like the Microsoft Kinect or advanced lidar sensors. 3D vision enables robots to interact with objects in a more sophisticated manner, such as picking up objects with different shapes and sizes or navigating through complex environments with obstacles.

Furthermore, object recognition is not limited to static objects. Robots equipped with mastery in this field can also detect and predict the movements of dynamic objects, such as people or vehicles. This capability is crucial for robots operating in dynamic environments like busy city streets or crowded airports, where anticipating the behavior of moving objects is essential for safe and efficient navigation.

Robotic vision and object recognition mastery have far-reaching implications across various domains. In healthcare, robots can assist surgeons by recognizing surgical instruments or monitoring a patient's condition. In manufacturing, robots can identify defects in products and ensure quality control. In agriculture, they can recognize ripe fruits for harvesting. In the retail industry, robots can assist customers by locating products on shelves. These applications highlight the versatility and impact of robotic vision and object recognition.

However, challenges remain in achieving true mastery in this field. Robotic vision needs to work robustly in various lighting conditions, handle occlusions (when objects partially block each other), and adapt to diverse environments. Furthermore, ensuring the privacy and security of visual data is a growing concern, as robots equipped with cameras may inadvertently capture sensitive information.

In summary, robotic vision and object recognition mastery represent a transformative capability in robotics. It empowers robots to understand and interact with their environment, recognize objects, anticipate movements, and contribute to various fields such as healthcare, manufacturing, agriculture, and retail. With continued advancements in deep learning and 3D perception, the future of robotic vision holds the promise of even greater versatility and precision. As robots become increasingly capable in this area, they will undoubtedly become more valuable partners in our everyday lives.

Chapter 5: Advanced Robot Kinematics and Dynamics

Exploring the realm of advanced kinematics solutions for manipulators, we delve deeper into the intricate mechanics that enable robots to perform precise and complex tasks. It's akin to understanding the graceful dance of a robotic arm as it navigates through its workspace, positioning itself with remarkable precision. These advanced kinematics solutions represent a cornerstone of robotics, enabling robots to achieve extraordinary dexterity and versatility in a wide range of applications.

At its core, kinematics is the study of motion, and in the context of robotics, it deals with the motion of robot arms, commonly referred to as manipulators. These robotic arms are typically composed of a series of connected segments, known as links, and joints that allow movement between these links. Kinematics is concerned with calculating the positions, orientations, and trajectories of the end-effector (the tool or hand) of the robotic arm based on the movements of its joints.

Now, let's journey into some of the advanced kinematic solutions that propel robotics into uncharted territory. One of the fundamental aspects is forward kinematics, which involves determining the end-effector's position and orientation in the robot's workspace for a given set of joint angles. Think of it as the ability to answer the question, "Where is the robot's hand right now?" This knowledge is critical for tasks that require precise positioning, such as pick-and-place operations in manufacturing or surgical procedures in healthcare.

Inverse kinematics, on the other hand, is like solving a puzzle in reverse. It deals with calculating the joint angles required to achieve a specific position and orientation of the end-effector. This is particularly useful when you know where you want the robot's hand to be, and you need to figure out how to get there efficiently. Inverse kinematics plays a vital role in tasks like robotic arm control and path planning.

Singularities are another intriguing aspect of advanced kinematics. These are configurations where the robot's motion becomes highly sensitive or degenerate. Imagine a situation where a robotic arm cannot move smoothly, and its movement suddenly becomes erratic or even impossible. Singularities are like the potholes in the road of robotics, and understanding them is crucial for avoiding issues in real-world applications.

Redundancy resolution is a captivating concept in advanced kinematics, especially in the realm of redundant manipulators. These are robotic arms with more joints than strictly necessary for a task. Redundancy can be a tremendous advantage, as it provides flexibility and adaptability in various situations. However, it also introduces challenges, such as determining how to control the extra joints effectively. Advanced kinematic solutions are essential for optimizing the use of redundant degrees of freedom.

Another exciting frontier in kinematics is parallel manipulators. These are robotic systems where multiple limbs or chains work in parallel to control a common end-effector. Picture a spider-like robot with multiple legs moving in coordination. Parallel manipulators offer advantages in terms of stability, precision, and speed, making them ideal for applications like flight simulators, 3D printing, or high-precision machining.

Now, let's consider the practical applications of these advanced kinematic solutions. In the world of industrial robotics, advanced kinematics enable robots to perform tasks with unparalleled accuracy. For instance, in automotive manufacturing, robots equipped with advanced kinematics can precisely weld, assemble, and paint car components. This level of precision is critical for ensuring product quality and consistency.

In the field of medical robotics, advanced kinematics solutions are invaluable. Robotic surgical systems rely on kinematic calculations to enable surgeons to perform minimally invasive procedures with remarkable precision. These robots can navigate through intricate anatomical structures, making surgeries less invasive, reducing patient trauma, and speeding up recovery times.

Agriculture is yet another domain where advanced kinematics are making a significant impact. Robots with advanced kinematic capabilities can harvest crops with delicacy and precision, ensuring minimal damage to plants and produce. This is particularly crucial in industries such as vineyard management or fruit picking.

Moreover, kinematics plays a vital role in the emerging field of service robotics. Robots designed to assist people in their daily lives need to be able to manipulate objects, open doors, or even cook meals. Advanced kinematic solutions enable these robots to interact with their environment and perform tasks that require a high level of dexterity.

In essence, advanced kinematics solutions for manipulators are the engine that drives the versatility and precision of robotic systems across various industries. Whether it's the smooth movements of a surgical robot, the precise assembly of electronics, or the delicate harvesting of crops, advanced kinematics enable robots to excel in tasks that were once considered impossible. As we continue to push the boundaries of robotics, the mastery of kinematics will remain a cornerstone of innovation, opening up new possibilities and transforming the way we live and work with robotic companions.

Optimization techniques in robot motion planning are like the wizardry that helps robots find the most efficient and effective paths through complex environments. Imagine a robot autonomously navigating a cluttered room, smoothly avoiding obstacles, and reaching its destination with grace and precision. These techniques form the backbone of intelligent robot behavior, enabling them to move in the real world with finesse.

To truly appreciate the importance of optimization in robot motion planning, let's first understand the complexities robots face. In a dynamic and uncertain environment, robots encounter numerous obstacles, tight spaces, and diverse terrain. They must plan their movements to avoid collisions, optimize energy consumption, and minimize travel time—all while considering their physical limitations.

One of the fundamental concepts in optimization for robot motion planning is the cost function. This function assigns a value to each possible path a robot can take. The robot's goal is to find the path with the lowest cost, which typically represents the shortest distance, the least energy consumption, or the safest route. The cost function encapsulates the robot's objectives and constraints, such as avoiding collisions or staying within its mechanical limits.

Now, let's delve into some of the optimization techniques used in robot motion planning. A classic method is Dijkstra's algorithm, which calculates the shortest path in a graph. This algorithm is like a cartographer tracing the quickest route on a map, considering all possible paths to reach the destination. However, Dijkstra's algorithm is computationally expensive, especially in large environments, and it doesn't account for obstacles.

To tackle obstacle avoidance, we turn to the A* algorithm, a heuristic search algorithm that combines the best of both worlds. It considers both the cost to reach a point and a heuristic estimate of the remaining cost to the goal. Think of it as a GPS that not only knows your current location and destination but also predicts the traffic conditions ahead. A* efficiently finds paths that avoid obstacles while optimizing for different criteria.

Another optimization gem in robot motion planning is the Rapidly-Exploring Random Tree (RRT) algorithm. RRT is like a curious explorer navigating a dark forest. It incrementally builds a tree of possible robot states by randomly sampling the configuration space and connecting them to the existing tree. This method is particularly effective in high-dimensional spaces and can quickly find feasible paths through cluttered environments.

For robotic systems requiring continuous motion, the use of optimization techniques like gradient-based methods comes into play. These methods involve iteratively adjusting the robot's path to minimize a cost function, often represented by a mathematical equation. Imagine a robot fine-tuning its trajectory as it moves, like a dancer refining their steps during a performance.

In the realm of optimization, there's also a subset known as constrained optimization. This is like solving a puzzle with specific rules. Robots often have constraints imposed on their movements, such as limited joint angles, payload capacities, or energy resources. Constrained optimization techniques ensure that robot motions adhere to these limitations while still finding optimal paths.

Incorporating uncertainty is essential in real-world robot motion planning. After all, the world is not always predictable. Probabilistic methods like Monte Carlo Localization or Markov Decision Processes (MDPs) consider uncertainty when making decisions. MDPs are particularly fascinating, as they model the robot's interaction with its environment as a series of decisions, factoring in the uncertainty of outcomes, similar to a chess player thinking several moves ahead.

Trajectory optimization is another vital aspect of robot motion planning. Instead of planning a single path, robots can optimize entire trajectories, accounting for dynamics, control inputs, and external forces. This is like a gymnast planning every twist and turn before executing a flawless routine, considering every muscle movement and torque applied.

Machine learning techniques have also made their mark in optimization for robot motion planning. Reinforcement learning, for instance, allows robots to learn optimal policies through trial and error. It's like a robot playing a game of chess and learning the best moves after multiple matches. Once trained, these robots can adapt to various environments and situations.

Human-robot interaction (HRI) benefits greatly from optimization techniques. Robots operating in close proximity to humans must not only plan efficient paths but also consider social norms and comfort. Optimization methods can be used to find paths that respect personal space, follow social conventions, and ensure a pleasant coexistence with humans.

In summary, optimization techniques in robot motion planning are the secret sauce that allows robots to move with grace, intelligence, and efficiency in the real world. From Dijkstra's algorithm mapping the shortest route to A* steering clear of obstacles, from RRT exploring new territories to gradient-based methods fine-tuning continuous motion, optimization is the compass that guides robots through the maze of our dynamic world. Constrained optimization, probabilistic methods, and trajectory optimization provide the essential flexibility and adaptability required for complex tasks. Machine learning and reinforcement learning add a layer of adaptiveness and autonomy, making robots smarter with experience. Finally, in the realm of HRI, optimization ensures that robots not only navigate effectively but also interact harmoniously with humans. As we continue to advance these optimization techniques, robots will become even more proficient, seamlessly integrating into our daily lives, and transforming industries with their capabilities.

Chapter 6: Expert-Level Robot Control and Navigation

Advanced control techniques are the high-level strategies that empower robots to perform complex tasks with precision and adaptability. They're like the conductor's baton in an orchestra, orchestrating a symphony of robotic movements to achieve intricate goals. These techniques go beyond basic control and are essential for applications in manufacturing, autonomous vehicles, medical robotics, and more.

One of the fundamental concepts in advanced control techniques is feedback control. It's akin to a thermostat maintaining a comfortable room temperature. In robotics, sensors constantly measure the robot's state, such as position, velocity, and orientation. The controller then uses this information to make real-time adjustments to achieve desired outcomes. This loop of sensing and adjusting ensures accuracy and stability in various robotic tasks.

But let's dive deeper into some of the sophisticated control strategies used in robotics. Nonlinear control techniques are like a chef crafting a gourmet dish with complex flavors. In many robotic systems, the relationships between inputs and outputs are nonlinear, meaning they don't follow a simple linear pattern. Nonlinear control methods allow robots to handle these complex relationships, making them suitable for tasks like agile maneuvering and precise manipulation.

Adaptive control techniques are all about robots learning from experience, much like humans adapting to new situations. These controllers adjust their behavior based on the robot's interactions with the environment. For instance, in a robotic arm assembling products on a conveyor belt, adaptive control ensures that the robot can handle variations in product size and weight, continuously improving its performance.

Model-based control techniques are akin to a pilot using a flight simulator to practice before flying a real aircraft. Robots can have detailed mathematical models that describe their dynamics and behavior. These models are used to predict how the robot will respond to different control inputs. By using this foresight, robots can execute complex tasks with high precision, like a surgical robot performing delicate procedures.

In the world of robotics, there's often a need for robustness, much like a gymnast maintaining balance on a wobbly beam. Robust control techniques ensure that robots can perform reliably in the face of uncertainty and disturbances. These methods account for variations in the environment or the robot's condition, enabling it to maintain stability and complete tasks even when unexpected events occur.

For tasks that involve multiple robots working together, coordination and synchronization become crucial. Multi-agent control techniques are like choreographing a dance with multiple dancers, ensuring that all robots move harmoniously towards a common goal. Applications range from swarming drones to autonomous vehicles navigating urban traffic.

Optimal control techniques aim to find the best possible actions for a robot to achieve its goals while considering various constraints. It's like a chess grandmaster calculating every move to reach checkmate. Optimal control algorithms help robots make decisions that optimize criteria such as energy consumption, time efficiency, or safety. These techniques are essential in fields like robotics for space exploration or autonomous driving.

For robots to interact safely with humans, control techniques that ensure safety are paramount. Safety control methods are like the guardrails on a highway, preventing accidents and harm. In collaborative robotics, where robots work alongside humans, safety control strategies are designed to detect and respond to potentially dangerous situations to protect both humans and robots.

In the realm of artificial intelligence, reinforcement learning and deep reinforcement learning are gaining traction in advanced control techniques. These approaches enable robots to learn from their actions and experiences, much like how humans learn from trial and error. Reinforcement learning is employed in various applications, from robot sports to autonomous navigation, where robots adapt and improve their behavior over time. Predictive control is akin to a chess player thinking several moves ahead. In this method, robots use models to predict future states and optimize their actions accordingly. This approach is particularly valuable in dynamic environments, where robots need to make fast decisions to avoid collisions or respond to rapidly changing conditions.

In summary, advanced control techniques for robotics encompass a wide array of strategies that enable robots to perform intricate tasks with precision and adaptability. These techniques range from handling nonlinear dynamics to adapting to changing environments, learning from experience, ensuring safety, and optimizing actions for various criteria. As robotics continues to advance, these control techniques will play a pivotal role in shaping the capabilities of robots across various industries and applications. Navigating through an environment autonomously is a defining capability for many robots, akin to a human driver skillfully navigating a busy city. In the world of robotics, autonomous navigation involves robots making decisions in real-time to plan routes, avoid obstacles, and reach their destinations safely. This fundamental aspect of robotics is essential for various applications, from self-driving cars and drones to warehouse robots and space exploration rovers. At the heart of autonomous navigation is the ability to perceive the robot's surroundings, much like how our eyes and ears provide us with information about the environment. Sensors play a critical role here, and robots often use a combination of cameras, LiDAR, radar, ultrasonic sensors, and GPS to gather data about their surroundings. These sensors help robots build a map of the environment, detect obstacles, and determine their own position within it.

Simultaneous Localization and Mapping (SLAM) is a key technology in autonomous navigation, analogous to a traveler creating a map as they explore an unfamiliar city. SLAM algorithms enable a robot to create a map of its surroundings while simultaneously determining its own position within that map. This real-time mapping and localization are essential for a robot to navigate effectively in an unknown or dynamic environment.

Path planning is another crucial component of autonomous navigation. It's like a GPS system calculating the best route to a destination. In this context, robots use algorithms to find safe and efficient paths through the environment while avoiding obstacles. Path planning algorithms take into account factors such as the robot's current location, the desired destination, the map of the environment, and any dynamic obstacles that might be in the way.

Once a path is planned, the robot must execute it accurately. This involves control algorithms that adjust the robot's movements in real-time to follow the planned path. These algorithms are responsible for adjusting the robot's speed, steering, and braking, much like a driver adjusting the car's controls to stay on course.

Autonomous navigation is not limited to simple point-to-point movement. Robots often need to perform more complex maneuvers, similar to a skilled pilot landing an aircraft. For example, an autonomous car may need to merge onto a highway, change lanes, or navigate through a busy intersection. These advanced maneuvers require sophisticated decision-making algorithms that take into account the behavior of other vehicles, traffic rules, and safety considerations.

Decision-making in autonomous navigation is further complicated by uncertainty and the need to react to unexpected events. For example, a self-driving car must be able to respond to sudden changes in traffic conditions, such as a vehicle swerving into its lane or a pedestrian crossing the road unexpectedly. Decision-making algorithms use a combination of sensor data, environmental models, and predefined rules to make split-second decisions in such situations.

Machine learning and artificial intelligence (AI) are playing an increasingly important role in autonomous navigation. These technologies allow robots to learn from data and improve their navigation capabilities over time, much like a human driver becomes more skilled with experience. Machine learning can be used for tasks such as recognizing objects and predicting the behavior of other agents in the environment.

Human-robot interaction is another critical aspect of autonomous navigation, especially in scenarios where robots share spaces with humans. Ensuring the safety of both humans and robots requires sophisticated algorithms for detecting and responding to human presence and intent. This is particularly important in settings like autonomous delivery robots in urban areas or collaborative robots (cobots) working alongside human workers in factories. Autonomous navigation is not limited to terrestrial robots. It extends to aerial and marine robots as well. For example, autonomous drones use similar principles to navigate through the air, avoid obstacles, and perform tasks such as surveillance, search and rescue, or package delivery. Autonomous underwater vehicles (AUVs) use specialized sensors and algorithms to explore the ocean depths and gather scientific data. In summary, autonomous navigation and decision-making are foundational capabilities for a wide range of robots operating in diverse environments. These robots rely on sensors, mapping, path planning, control algorithms, decision-making, machine learning, and human-robot interaction to navigate autonomously and accomplish their tasks safely and effectively. As technology continues to advance, the field of autonomous navigation holds the promise of revolutionizing transportation, logistics, exploration, and many other industries.

Chapter 7: Cutting-Edge AI and Machine Learning in Robotics

Deep learning, a subfield of artificial intelligence (AI), has ushered in a transformative era in robotics by significantly enhancing the capabilities of robotic perception. Imagine a robot that can see, hear, and understand its surroundings with remarkable precision, much like the way our own senses help us make sense of the world. Deep learning has enabled robots to achieve this level of perception, opening up new possibilities across various domains, from manufacturing and healthcare to exploration and entertainment.

At the core of deep learning's impact on robotic perception are neural networks. These computational models are inspired by the structure of the human brain, composed of interconnected nodes or "neurons" that process information. Neural networks excel at tasks that involve recognizing patterns and making predictions, making them ideal for tasks like image and speech recognition—key components of robotic perception.

For robots, visual perception is of paramount importance. Just as our eyes capture images of the world, cameras on robots capture images of their environment. Deep learning-based computer vision models enable robots to not only "see" but also understand what they see. Convolutional Neural Networks (CNNs), for instance, have proven highly effective at recognizing objects and people in images and videos. This capability is crucial for tasks like autonomous navigation, where robots must detect and avoid obstacles.

Furthermore, deep learning allows robots to go beyond basic object recognition and delve into more complex visual tasks. For instance, robots can use deep learning to estimate the pose and orientation of objects, making it possible to pick them up and manipulate them with dexterity. This is incredibly valuable in industries such as manufacturing, where robots work alongside humans on factory floors.

In addition to visual perception, deep learning has revolutionized robotic speech recognition and natural language processing. Imagine a robot that can listen to and understand spoken commands, allowing for intuitive human-robot interaction. Thanks to deep learning models like Recurrent Neural Networks (RNNs) and Transformers, robots can process and generate human language, making them more accessible and user-friendly.

Deep learning has also made it possible for robots to perform sentiment analysis on text and speech. This means they can gauge human emotions and respond accordingly. This capability has applications not only in customer service robots but also in healthcare, where robots can provide companionship and support to patients.

Another area where deep learning has had a profound impact is in the realm of autonomous vehicles. Self-driving cars, equipped with sensors and cameras, rely on deep neural networks for tasks like lane detection, object recognition, and decision-making. These networks process vast amounts of data in real-time, allowing the vehicle to navigate complex traffic scenarios safely.

Robotic perception isn't limited to the visible spectrum; it extends to other modalities as well. For instance, robots equipped with LiDAR (Light Detection and Ranging) sensors can use deep learning to process the 3D point cloud data generated by these sensors. This enables them to build detailed maps of their environment and detect obstacles with high accuracy.

Deep learning has also found applications in medical robotics, where it aids in tasks such as image-guided surgery and medical image analysis. Robots equipped with deep learning-based systems can assist surgeons by providing real-time feedback and enhancing precision during surgical procedures.

In the field of agricultural robotics, deep learning has been used to identify and classify plants, pests, and diseases. This technology allows robots to autonomously manage crops and optimize agricultural practices, contributing to increased crop yields and sustainability.

Robotic perception powered by deep learning has a profound impact on search and rescue operations. Robots equipped with advanced vision systems can navigate disaster-stricken areas, locate survivors, and identify hazards, all while providing crucial information to first responders.

However, it's worth noting that deep learning for robotic perception isn't without challenges. One significant challenge is the need for large amounts of labeled training data. Neural networks require extensive datasets to learn effectively, and curating such datasets can be time-consuming and expensive. Additionally, deep learning models can be computationally intensive, necessitating powerful hardware for real-time processing.

Furthermore, deep learning models may exhibit limitations in handling rare or unexpected scenarios. Robots operating in dynamic, unstructured environments must be able to adapt to novel situations, which can be a complex task for pre-trained models.

Despite these challenges, deep learning has undoubtedly revolutionized robotic perception. Its ability to process vast amounts of data and make predictions with remarkable accuracy has opened up new possibilities for robots across various industries. With continued advancements in deep learning techniques and hardware capabilities, the future holds even more exciting prospects for robotic perception, paving the way for robots that can interact with the world and humans in increasingly sophisticated ways.

Reinforcement learning, a prominent subfield of artificial intelligence, has emerged as a powerful technique for enabling autonomous robots to learn and adapt to their environments. Imagine a robot that can explore, make decisions, and improve its performance through trial and error, much like how we learn from our experiences. Reinforcement learning makes this a reality, offering robots the ability to acquire complex skills and make intelligent decisions in dynamic and uncertain settings.

At its core, reinforcement learning is a type of machine learning where an agent, in this case, a robot, interacts with an environment to achieve a goal. The agent learns by taking actions and receiving feedback in the form of rewards or penalties, much like a pet learning to perform tricks with treats as rewards. Over time, the robot learns to take actions that maximize cumulative rewards.

One of the key components of reinforcement learning is the Markov Decision Process (MDP), which is a mathematical framework used to model decision-making in an environment. In an MDP, the robot considers its current state, possible actions it can take, and the rewards associated with those actions. This allows the robot to choose actions that lead to the best expected outcomes.

A fundamental concept in reinforcement learning is the exploration-exploitation trade-off. This means that the robot must balance between exploring new actions to discover better strategies and exploiting known actions that have yielded positive outcomes. Striking the right balance is crucial for efficient learning, as excessive exploration can lead to slow progress, while excessive exploitation may lead to suboptimal solutions.

Reinforcement learning algorithms can be broadly categorized into model-free and model-based approaches. Model-free algorithms, such as Q-learning and deep reinforcement learning, directly learn the optimal policy (sequence of actions) without explicitly modeling the environment. Deep reinforcement learning, in particular, has gained prominence due to its success in training agents to perform tasks that require high-dimensional input, such as image-based perception.

Model-based approaches, on the other hand, build an explicit model of the environment, which the robot uses to plan and make decisions. These models can be useful in scenarios where exploration is costly or risky, as they allow the robot to simulate various actions and their consequences before taking them in the real world.

Reinforcement learning has found numerous applications in robotics. One of the most notable examples is autonomous navigation. Robots can learn to navigate complex environments, avoid obstacles, and reach specific destinations through reinforcement learning. This is particularly valuable in scenarios like warehouse automation and delivery robots.

Another exciting application is robotic manipulation. Reinforcement learning enables robots to learn how to grasp and manipulate objects with dexterity. Robots can learn the necessary motor skills and control policies to perform tasks like picking and placing objects, assembly, and even playing games like table tennis.

Reinforcement learning also plays a crucial role in swarm robotics, where groups of robots collaborate to achieve a common goal. These robots can learn to coordinate their actions and optimize their collective behavior through reinforcement learning algorithms.

Moreover, reinforcement learning has applications in robotic learning from human demonstrations. Robots can learn tasks by observing and imitating human actions, making them more adaptable and capable of assisting humans in various tasks.

Despite its promise, reinforcement learning in robotics is not without challenges. One significant challenge is the need for a large number of trials for the robot to learn effectively. In real-world scenarios, this can be impractical or expensive. Techniques like transfer learning and curriculum learning aim to mitigate this issue by leveraging knowledge from related tasks or gradually increasing the task complexity.

Additionally, reinforcement learning often requires high computational resources and may struggle with sample inefficiency. Robots need to interact with the environment to collect data, which can be time-consuming and resource-intensive.

Moreover, ensuring the safety of autonomous robots during the learning process is a critical concern. Reinforcement learning algorithms can potentially lead to unintended and unsafe behavior, necessitating the development of robust safety mechanisms and reward shaping techniques.

In summary, reinforcement learning is a promising approach for achieving autonomous robots that can learn and adapt to their environments. It has demonstrated its potential in various applications, from navigation and manipulation to swarm robotics and human-robot interaction. While challenges remain, ongoing research and technological advancements are paving the way for increasingly capable and intelligent autonomous robots, bringing us closer to a future where robots are valuable partners in a wide range of tasks and activities.

Chapter 8: Robotic Vision and Object Recognition Mastery

Advanced image processing is a critical component of robotic vision, enabling robots to perceive and understand their surroundings with a level of sophistication that approaches human vision. Robots equipped with advanced image processing capabilities can interpret and make sense of visual data, which is essential for tasks such as navigation, object recognition, and scene understanding.

One of the fundamental aspects of advanced image processing is image enhancement. This technique aims to improve the quality of images captured by robotic cameras, often in challenging environments with varying lighting conditions. By enhancing the contrast, brightness, and sharpness of images, robots can extract more information from the visual data they receive.

In addition to enhancement, image filtering is another essential tool in the toolbox of advanced image processing. Filtering allows robots to selectively emphasize or suppress certain features in an image, depending on the specific task. For example, edge detection filters highlight the boundaries between objects, making it easier for a robot to identify objects in a cluttered scene.

Robotic vision often involves the analysis of color images, and color correction is a crucial step in image processing. Color correction techniques ensure that the colors in images are accurate and consistent, despite variations in lighting conditions. This is particularly important in applications like object recognition and tracking.

Segmentation is another vital concept in advanced image processing. It involves dividing an image into meaningful regions or segments, allowing a robot to identify and analyze individual objects or regions of interest within a scene. Segmentation is a fundamental step in tasks like autonomous driving, where a robot must distinguish between the road, vehicles, pedestrians, and other objects.

Once images are enhanced, filtered, and segmented, robots can proceed to feature extraction. Feature extraction involves identifying specific patterns, textures, or characteristics within an image that are relevant to the task at hand. For instance, in facial recognition, robots extract features like the eyes, nose, and mouth to identify individuals.

Object recognition is one of the most prominent applications of advanced image processing in robotics. Robots equipped with object recognition algorithms can identify and classify objects in their environment. This capability is essential in tasks like industrial automation, where robots need to locate and handle specific objects on a production line.

Object tracking is another critical aspect of robotic vision. Robots with tracking capabilities can follow the movement of objects within their field of view. This is valuable in applications such as surveillance, where a robot needs to monitor and track the movement of people or vehicles.

Stereo vision is a more advanced technique in robotic vision that involves using multiple cameras to capture a three-dimensional view of the environment. By analyzing the disparities between images captured by different cameras, robots can estimate the depth and distance of objects in the scene. Stereo vision is crucial for tasks like 3D mapping and obstacle avoidance.

For robots to interact effectively with humans and their environment, they need to understand not only objects but also human gestures and expressions. Image processing techniques can be applied to recognize and interpret gestures and facial expressions, enabling robots to engage in natural and intuitive human-robot interactions.

Furthermore, advanced image processing plays a pivotal role in visual simultaneous localization and mapping (SLAM), a technique used by robots to navigate and create maps of unknown environments. By analyzing the visual data obtained from cameras, robots can simultaneously determine their position and map the surrounding environment.

Convolutional neural networks (CNNs) are a class of deep learning models that have revolutionized robotic vision. CNNs are designed to automatically learn and extract hierarchical features from images. They excel in tasks such as image classification, object detection, and semantic segmentation, making them indispensable for modern robotic vision systems.

Robotic vision is not limited to static images. Video processing is equally essential, enabling robots to analyze video streams in real-time. Real-time video processing is crucial in applications like autonomous driving, where a robot must continuously interpret the dynamic environment and make split-second decisions.

In summary, advanced image processing is a cornerstone of robotic vision, enabling robots to perceive and interact with the world in ways that were once the realm of science fiction. From enhancing and filtering images to segmenting, recognizing, and tracking objects, these techniques empower robots with the ability to navigate, manipulate, and understand their environments. As robotics continues to advance, the role of advanced image processing in shaping the capabilities of robots is only set to grow, ushering in a future where robots can seamlessly integrate into various aspects of our lives.

Object detection and tracking represent a critical frontier in the field of computer vision and robotics, allowing machines to perceive, identify, and follow objects or entities within their visual field. As technology continues to evolve, achieving expertise in object detection and tracking is essential for robots to interact effectively with their environments, perform tasks, and even collaborate with humans.

The foundation of object detection and tracking is rooted in the analysis of visual data. Robots equipped with cameras or other sensors capture images or video frames that serve as the raw input for these processes. These visual inputs are essentially snapshots of the robot's surroundings, providing a rich source of information for understanding and interacting with the world.

In the realm of object detection, the primary goal is to locate and identify specific objects or entities within the visual data. This process involves recognizing patterns, shapes, and features that correspond to known objects. For example, in autonomous vehicles, object detection algorithms can identify pedestrians, vehicles, traffic signs, and other objects on the road, contributing to safe navigation.

To achieve expert-level object detection, robots often employ deep learning techniques, particularly convolutional neural networks (CNNs). CNNs are a class of artificial neural networks designed to automatically learn and extract hierarchical features from images. They have shown remarkable performance in object detection tasks, thanks to their ability to discern complex patterns and variations in visual data.

A critical aspect of object detection is the ability to differentiate between objects in different categories. For instance, in a factory setting, robots must be capable of distinguishing between various product types on a conveyor belt. This level of expertise ensures that each product is handled appropriately and can contribute to efficient manufacturing processes.

Real-time object detection is another significant challenge in robotics. Robots often operate in dynamic environments where objects can appear, move, or disappear quickly. Achieving real-time object detection requires algorithms and hardware capable of processing visual data with low latency, enabling robots to respond swiftly to changing situations.

Once objects are detected, the next step is object tracking, which involves following the movement of these objects over time. Object tracking is a fundamental skill for robots in various applications, including surveillance, robotics-assisted surgery, and human-robot interaction. Tracking is particularly useful when objects move in and out of the robot's field of view, ensuring that they remain under continuous observation.

Object tracking algorithms rely on various techniques, such as feature-based tracking, where distinctive visual features of an object are used for tracking, and appearance-based tracking, where the overall appearance of an object is considered. These methods often combine aspects of machine learning and computer vision to estimate the object's position and motion accurately.

Expert-level object tracking involves overcoming challenges like occlusion, where objects may be partially or fully hidden from view, and abrupt changes in motion or appearance. Advanced tracking algorithms can handle these scenarios robustly, ensuring that the robot maintains a consistent understanding of the objects it tracks.

Multiple object tracking is another dimension of expertise in tracking. In scenarios with multiple objects, such as monitoring a crowd or managing a warehouse filled with moving items, robots must track and distinguish between multiple entities simultaneously. This requires sophisticated algorithms capable of handling object interactions, occlusions, and maintaining accurate identities for each object.

Beyond tracking individual objects, robots often engage in semantic tracking. Semantic tracking involves not only identifying objects but also understanding their roles and interactions within a scene. For example, in a kitchen environment, a robot must not only track objects like pots and pans but also recognize that these objects are used for cooking and have specific relationships with each other.

Human-robot interaction greatly benefits from expertise in object detection and tracking. Robots that can accurately perceive and track human movements and gestures can engage in natural and intuitive interactions with humans. This capability is essential in various domains, including healthcare, where robots assist with patient care, or in smart homes, where robots serve as personal assistants.

Expert-level object detection and tracking also find applications in the entertainment industry, where robots are used for filming and special effects. In these settings, robots need to precisely track the positions and movements of actors, props, and cameras to create engaging and realistic visual content.

In summary, object detection and tracking at an expert level are indispensable skills for robots operating in diverse domains. Achieving expertise in these areas empowers robots with the ability to navigate complex environments, interact with objects and humans effectively, and contribute to a wide range of applications. As robotics technology continues to advance, the quest for ever-improving object detection and tracking capabilities remains at the forefront of research and development, driving the field towards new frontiers of innovation.

Chapter 9: Leadership in Robotics Research and Innovation

Securing funding for robotics projects is a vital step in turning innovative ideas into reality. Whether you are an academic researcher, a startup entrepreneur, or a seasoned professional, obtaining the necessary financial resources to support your robotics endeavors can be a challenging yet rewarding journey.

The first step in the process of securing funding is to clearly define your project's goals and objectives. This involves articulating what problem your robotics project aims to solve, what specific outcomes you expect to achieve, and how your work will contribute to the broader field of robotics. A well-defined project proposal not only serves as a roadmap for your work but also communicates your vision and passion to potential funders.

Once you have a clear project proposal, the next step is to identify potential sources of funding. These sources can vary widely, depending on the nature of your project, your goals, and your organization's profile. Common sources of funding for robotics projects include government grants and contracts, private foundations, industry partnerships, venture capital, and crowdfunding.

Government grants and contracts are often available for robotics research and development projects, particularly in areas with strategic importance, such as defense, healthcare, and transportation. These funding opportunities can provide substantial financial support, but they typically come with rigorous application and reporting requirements. It's essential to carefully review the guidelines and eligibility criteria for government funding programs and tailor your proposal accordingly.

Private foundations and nonprofit organizations also offer funding opportunities for robotics projects, often with a focus on specific research areas or societal challenges. These organizations may have a strong interest in supporting projects that align with their mission and objectives. Building relationships with program officers and demonstrating a clear alignment between your project and the foundation's goals can enhance your chances of securing funding.

Industry partnerships can be an attractive option for robotics startups and companies looking to bring their innovations to market. Collaborating with industry leaders can provide not only financial support but also access to expertise, resources, and potential customers. When seeking industry partnerships, it's crucial to identify mutually beneficial opportunities and articulate how your project can address industry needs and challenges.

Venture capital is another funding avenue for robotics startups and entrepreneurs seeking to commercialize their technologies. Venture capitalists are often interested in high-growth opportunities with the potential for significant returns on investment. To attract venture capital funding, you need to develop a compelling business plan, demonstrate market demand for your robotics solution, and showcase a talented team capable of executing the project.

Crowdfunding platforms, such as Kickstarter and Indiegogo, have become increasingly popular for raising funds for robotics projects. These platforms allow individuals and small teams to showcase their projects to a global audience and solicit contributions from backers who believe in their ideas. Successful crowdfunding campaigns require effective storytelling, engaging visuals, and clear incentives for backers.

Grant writing and proposal development are essential skills when seeking funding for robotics projects. Crafting a persuasive proposal involves clearly articulating the significance of your project, detailing the research or development plan, outlining the budget, and explaining how the funding will be used to achieve specific milestones. It's essential to tailor your proposals to the priorities and requirements of each funding source.

Networking and relationship-building are also crucial components of the funding process. Establishing connections with potential funders, collaborators, and mentors within the robotics community can open doors to funding opportunities and provide valuable guidance and support. Attend conferences, workshops, and industry events to meet like-minded individuals and organizations that share your passion for robotics.

Securing funding for robotics projects often requires persistence and adaptability. Many successful researchers and entrepreneurs face rejection before finding the right funding opportunity. It's essential to learn from feedback, refine your proposals, and continue exploring diverse funding sources. Additionally, staying informed about emerging trends, technologies, and funding opportunities in the robotics field can give you a competitive edge.

In summary, securing funding for robotics projects is a multifaceted process that involves defining project goals, identifying funding sources, crafting persuasive proposals, networking, and adapting to feedback. Whether your goal is to advance the frontiers of robotics research, launch a robotics startup, or develop innovative solutions for real-world challenges, obtaining the necessary financial support is a crucial step towards bringing your vision to life. As you embark on your journey to secure funding, remember that perseverance, passion, and strategic thinking are your allies in the pursuit of your robotics dreams.

Mentoring the next generation of roboticists is a noble and rewarding endeavor that plays a pivotal role in the growth and development of the field of robotics. As experienced individuals who have walked the path of robotics research, development, or entrepreneurship, it is our responsibility to guide and inspire the budding minds eager to make their mark in this exciting and ever-evolving domain.

One of the first steps in mentoring future roboticists is to foster their curiosity and passion for robotics. Encouraging young individuals to explore their interests, tinker with robots or robotic components, and ask questions can ignite a spark of enthusiasm that may lead to a lifelong pursuit of robotics. A mentor's role is to provide guidance and resources that enable these young minds to discover the fascinating world of robotics.

As mentors, we should also emphasize the importance of a solid educational foundation in robotics. Robotics is an interdisciplinary field that draws from mathematics, physics, computer science, and engineering, among others. It is essential to guide aspiring roboticists toward educational programs or courses that will equip them with the necessary knowledge and skills. This might involve recommending specific universities, online courses, or textbooks that align with their interests and career goals.

Moreover, mentoring involves exposing mentees to a diverse range of robotics topics and applications. Robotics is not limited to a single domain but spans various areas such as industrial automation, healthcare robotics, autonomous vehicles, and more. Mentors can organize workshops, guest lectures, or hands-on projects to expose mentees to different facets of robotics, helping them discover their specific areas of interest.

Another crucial aspect of mentoring is providing practical experience and hands-on opportunities. Robotics is a field where learning by doing is highly effective. Mentors can collaborate with educational institutions, research labs, or robotics companies to create internships, research projects, or apprenticeships for aspiring roboticists. These real-world experiences allow mentees to apply their theoretical knowledge, develop practical skills, and gain insights into the challenges and opportunities in the field.

Additionally, mentoring involves nurturing essential soft skills that are often overlooked but are equally important in a roboticist's journey. Effective communication, teamwork, problem-solving, and adaptability are skills that can significantly impact a roboticist's success. Mentors can encourage mentees to participate in group projects, presentations, and collaborations to develop these skills.

Mentoring also entails providing guidance on setting clear goals and objectives. Robotics is a vast field with numerous research directions and career paths. Mentors can help mentees define their short-term and long-term goals, whether it's pursuing a Ph.D., developing a robotics startup, or contributing to open-source robotics projects. Setting goals provides a sense of direction and motivation for future roboticists.

Furthermore, mentors should instill a sense of ethics and responsibility in their mentees. Robotics has the potential to create transformative technologies, but it also raises ethical and societal questions. Mentors can encourage their mentees to consider the ethical implications of their work, emphasizing the importance of responsible and ethical robotics research and development.

In addition to sharing knowledge and experiences, mentors should be approachable and supportive. Building a trusting and open mentor-mentee relationship is essential for effective guidance. Mentees should feel comfortable asking questions, seeking advice, and sharing their aspirations and concerns. Being a mentor means being available to offer encouragement, listen to mentees' ideas, and provide constructive feedback.

Mentors should also promote diversity and inclusivity in the field of robotics. Encouraging individuals from underrepresented backgrounds, including women and minorities, to pursue careers in robotics is crucial for the field's growth and innovation. Mentors can actively engage in initiatives that promote diversity and provide mentorship opportunities for aspiring roboticists from diverse backgrounds. As robotics mentors, we should stay up-to-date with the latest developments in the field. Robotics is a dynamic and rapidly evolving domain, with breakthroughs and advancements occurring regularly. Staying informed about emerging technologies, trends, and research areas enables mentors to guide their mentees effectively and expose them to cutting-edge ideas and opportunities.

In summary, mentoring the next generation of roboticists is a valuable contribution to the field's growth and sustainability. Mentors play a pivotal role in inspiring, educating, and guiding young minds who aspire to make a meaningful impact in robotics. By fostering curiosity, providing educational resources, offering practical experiences, nurturing soft skills, setting goals, promoting ethics, and supporting diversity, mentors can empower aspiring roboticists to thrive in this exciting and multidisciplinary field. Ultimately, mentoring is a collaborative and rewarding journey that shapes the future of robotics.

Chapter 10: The Future of Robotics Research: Pioneering New Frontiers

Emerging technologies and trends in robotics are continually shaping the landscape of this dynamic and rapidly evolving field. Robotics, once confined to the realm of science fiction, has become an integral part of various industries and everyday life, thanks to ongoing advancements and innovative developments. In this chapter, we will explore some of the most exciting and transformative technologies and trends that are propelling robotics into the future.

One of the most prominent trends in robotics is the rise of collaborative robots, commonly referred to as cobots. Unlike traditional industrial robots, cobots are designed to work alongside humans in a shared workspace. These robots are equipped with advanced sensors and safety features that allow them to operate safely alongside human workers. Cobots are revolutionizing manufacturing, logistics, and healthcare by enhancing efficiency and productivity while reducing the risk of workplace accidents.

Artificial intelligence (AI) and machine learning have become indispensable tools in the robotics domain. Machine learning algorithms enable robots to perceive and interact with their environments in increasingly sophisticated ways. Computer vision, natural language processing, and reinforcement learning are just a few examples of AI techniques that empower robots to understand and respond to the world around them. As AI continues to advance, robots are becoming more adaptable, autonomous, and capable of handling complex tasks.

Another notable trend is the proliferation of robots in healthcare and medicine. Robots are being used for surgery, diagnostics, rehabilitation, and even patient care. Surgical robots, for instance, enable surgeons to perform minimally invasive procedures with greater precision, resulting in faster recovery times for patients. Telemedicine robots allow healthcare professionals to remotely examine patients and provide medical care, bridging geographical gaps and improving access to healthcare services. Robotics is also making significant strides in the field of autonomous vehicles. Self-driving cars and drones equipped with autonomous navigation systems are poised to revolutionize transportation and logistics. These vehicles rely on a combination of sensors, GPS, and AI algorithms to navigate and make decisions in real-time. The development of autonomous vehicles has the potential to reduce traffic accidents, decrease fuel consumption, and transform the way goods are delivered. The integration of robots into smart homes and smart cities is another compelling trend. Household robots, such as robotic vacuum cleaners and personal assistants, are becoming more sophisticated and capable of handling various household tasks. In smart cities, robots are being deployed for tasks like monitoring environmental conditions, delivering packages, and providing public services. These technologies have the potential to improve urban living, enhance sustainability, and optimize resource management. Robotics is also advancing in the realm of space exploration. Robots and rovers are playing a crucial role in planetary exploration missions. For instance, the Mars rovers, including Curiosity and Perseverance, are equipped with advanced instruments to study the Martian terrain, search for signs of past life, and collect samples. These robotic explorers are expanding our understanding of the cosmos and paving the way for future human missions to other planets.

The field of soft robotics is gaining momentum as researchers seek to create robots with more flexibility and adaptability. Soft robots are constructed from pliable materials and are inspired by natural organisms like octopuses and caterpillars. These robots have applications in areas such as search and rescue, where their ability to squeeze through tight spaces and conform to irregular shapes is advantageous. Soft robotics also shows promise in medical devices and wearable technologies.

In addition to these trends, the convergence of robotics with other emerging technologies is opening up new possibilities. For example, the fusion of robotics with 5G technology enables robots to communicate and operate with low latency and high reliability, making them more suitable for applications like remote surgery and telepresence. Blockchain technology is being explored to enhance the security and transparency of robotic systems, particularly in supply chain and logistics operations. Furthermore, biologically inspired robotics, often referred to as bio-robotics, draws inspiration from nature to develop robots that mimic the capabilities of living organisms. Examples include robots with insect-like locomotion, which are ideal for navigating challenging terrains, and drones that emulate the flight patterns of birds for improved maneuverability. In summary, robotics is experiencing a transformative era driven by a myriad of emerging technologies and trends. Collaborative robots, artificial intelligence, healthcare applications, autonomous vehicles, smart environments, space exploration, soft robotics, and interdisciplinary collaborations are all contributing to the rapid evolution of the field. The future of robotics holds tremendous promise, with the potential to revolutionize industries, enhance our daily lives, and expand our horizons in exploration and innovation. As we continue to explore and embrace these emerging technologies, the possibilities for robotics are limitless, and the journey is bound to be both exciting and enlightening. Robots have been pivotal in expanding our understanding of the cosmos and have played a crucial role in space exploration and colonization efforts. As we venture into the vast expanse of space, these robotic pioneers are indispensable, helping us overcome the challenges of exploring and settling in environments beyond Earth. Space exploration has always captured our imagination, and robots have been our eyes, ears, and hands in the far reaches of the universe. Starting with the lunar landers of the Apollo program, robots have been exploring celestial bodies, providing us with invaluable data about the moon, Mars, and beyond. Rovers like Sojourner, Spirit, Opportunity, and Curiosity have roamed the Martian surface, analyzing soil, rocks, and even searching for signs of past or present life. These robots are equipped with a wide array of scientific instruments, allowing them to carry out experiments and collect samples. They are truly our scientific ambassadors in the cosmos. Beyond Mars, robots have ventured even farther into our solar system. The Voyager probes, launched in the late 1970s, are still sending back data from interstellar space, providing insights into the outer reaches of our solar system. The Cassini-Huygens mission, which explored Saturn and its moon Titan, delivered stunning images and vital information about these distant worlds. Robotic missions to comets and asteroids have shed light on the formation of our solar system, and spacecraft like the New Horizons probe have given us our first close-up looks at Pluto and the Kuiper Belt. In the quest for understanding the cosmos, robots have become our eyes on the universe, capturing breathtaking images and helping us decipher the mysteries of the cosmos. The Hubble Space Telescope, equipped with a suite of instruments, has delivered awe-inspiring images of distant galaxies, nebulae, and planets. It has revolutionized our understanding of the universe's age, composition, and expansion. Upcoming space telescopes, like the James Webb Space Telescope, promise to expand our horizons even further, allowing us to peer deeper into space and time.

One of the most significant trends in space exploration is the development of autonomous robots capable of long-duration missions. As we set our sights on distant destinations, such as Mars and beyond, these robots are designed to operate independently for extended periods, making decisions based on their own observations and data analysis. The Perseverance rover, equipped with artificial intelligence, can autonomously navigate the Martian terrain, avoid hazards, and select interesting scientific targets, reducing the need for constant human intervention.

Colonizing other celestial bodies, like the Moon and Mars, is an ambitious goal that requires careful planning and the assistance of robotic systems. Before humans can establish a sustainable presence on these worlds, robots must pave the way by building habitats, extracting resources, and conducting experiments. NASA's Artemis program aims to return astronauts to the Moon by the mid-2020s, and robots will play a pivotal role in preparing for human missions. For example, the Resource Prospector mission was intended to send a rover to the Moon's south pole to search for water ice, a critical resource for future lunar colonies.

In the more distant future, Mars is a prime candidate for human colonization, and robots are essential precursors to human arrival. Robotic landers, rovers, and even flying drones will scout the Martian terrain, identifying suitable landing sites, testing life support systems, and conducting experiments. These robotic missions will be instrumental in ensuring the safety and success of future Martian settlers.

The development of in-situ resource utilization (ISRU) technologies is a key focus of robotic missions to celestial bodies. ISRU involves extracting and using local resources, such as water and minerals, to support human and robotic activities. On Mars, for example, robots may extract water from the Martian soil, generating oxygen and hydrogen for rocket fuel and life support systems. This approach reduces the need to transport these critical supplies from Earth, making long-duration missions and colonization more sustainable.

Robots are also contributing to the development of advanced propulsion technologies that could revolutionize space travel. Concepts like solar sails, ion propulsion, and nuclear thermal propulsion are being tested on robotic missions, paving the way for faster and more efficient interplanetary travel. These innovations will not only benefit robotic explorers but will also make crewed missions to distant destinations more feasible.

In the search for extraterrestrial life, robots have been our primary investigators. Missions to Mars, Europa (a moon of Jupiter), and Enceladus (a moon of Saturn) have focused on identifying habitable environments and searching for signs of life. Submersible robots may one day explore the subsurface oceans of icy moons, where conditions might be conducive to life. Robots equipped with mass spectrometers, microscopes, and other analytical instruments are poised to make groundbreaking discoveries if they encounter microbial life or organic compounds on other worlds. Another exciting frontier is the development of swarm robotics for space exploration. Swarms of small, interconnected robots can work together to accomplish complex tasks, such as mapping uncharted terrain, assembling structures, or exploring caves and tunnels. These swarms offer redundancy, adaptability, and the ability to tackle large-scale challenges efficiently. In summary, robotics is at the forefront of space exploration and colonization efforts, playing a pivotal role in expanding our knowledge of the universe and paving the way for human expansion beyond Earth. From rovers and landers on distant planets to autonomous spacecraft and resource utilization technologies, robots are our trailblazers in the cosmos. As we continue to push the boundaries of space exploration, we can look forward to even more remarkable discoveries and innovations driven by these robotic explorers.

In a rapidly changing world facing an array of complex global challenges, robotics has emerged as a versatile and dynamic tool with the potential to address critical issues on a planetary scale. From climate change and environmental sustainability to healthcare, disaster response, and food security, the role of robotics in tackling these global challenges is both promising and multifaceted.

One of the most pressing global challenges is climate change, driven primarily by the increase in greenhouse gas emissions. Robotics plays a crucial role in mitigating climate change through various applications. Autonomous drones and ground robots are used for environmental monitoring, helping scientists gather data on deforestation, ice melt, and carbon dioxide levels. These robots can access remote and hazardous areas, providing valuable information for policymakers and researchers. In agriculture, robots are aiding in precision farming, optimizing resource use, and reducing the environmental impact of agriculture. Robotic systems can precisely apply fertilizers and pesticides, reducing waste and preventing soil degradation. Additionally, robots are being deployed in renewable energy industries, such as wind and solar farms, for maintenance and inspection, increasing the efficiency and reliability of clean energy production.

Environmental sustainability extends beyond climate change, encompassing broader efforts to preserve and protect the natural world. Marine robotics, including autonomous underwater vehicles (AUVs) and remotely operated vehicles (ROVs), are crucial in exploring and conserving Earth's oceans. These robots help scientists study marine ecosystems, track ocean currents, and investigate the impacts of pollution and overfishing. In wildlife conservation, robotics aids in tracking and protecting endangered species. For example, drones equipped with thermal cameras can monitor wildlife populations, detect poaching activities, and assist in anti-poaching efforts.

Another formidable global challenge is healthcare, and robotics is making significant contributions to both medical treatment and healthcare accessibility. Surgical robots, like the da Vinci Surgical System, enable minimally invasive surgeries with precision and reduced recovery times. Telemedicine robots bring medical expertise to remote and underserved areas, connecting patients with healthcare providers through video conferencing and remote diagnostics. In the wake of the COVID-19 pandemic, robots have been used for tasks such as disinfection and patient care in hospitals to reduce the risk of infection. The development of robotic exoskeletons offers mobility solutions for people with disabilities, enhancing their quality of life. In disaster response and recovery, robots are invaluable assets that can save lives and mitigate damage. Unmanned aerial vehicles (UAVs), commonly known as drones, are rapidly deployed to assess disaster-stricken areas, helping first responders identify hazards, locate survivors, and plan rescue missions. Ground robots equipped with cameras, sensors, and manipulators can navigate through debris, search for survivors, and provide real-time data to rescue teams. These robots reduce the risk to human responders in hazardous environments and enhance the efficiency of disaster response efforts. In the aftermath of natural disasters, such as earthquakes and hurricanes, robotic systems can be employed for debris removal, structural inspections, and rebuilding processes.

Food security is a global concern exacerbated by factors such as population growth, climate change, and resource constraints. Robotics is revolutionizing agriculture and food production to address these challenges. Autonomous tractors and harvesters can operate around the clock, optimizing crop yields and reducing the labor required for farming. Robots equipped with advanced sensors and artificial intelligence can monitor crop health, detect diseases, and apply targeted interventions, reducing the need for chemical inputs. Indoor vertical farming, supported by robotics and automation, offers a sustainable solution for growing crops in urban environments, reducing transportation costs and emissions associated with traditional agriculture.

The global challenges of urbanization and transportation are also being addressed by robotics. Autonomous vehicles, including self-driving cars and drones, have the potential to revolutionize transportation systems, reducing traffic congestion, accidents, and greenhouse gas emissions. In densely populated urban areas, robots are being tested for last-mile delivery, providing a more efficient and environmentally friendly alternative to traditional delivery methods. These innovations have the potential to transform how people and goods move within and between cities, contributing to more sustainable and accessible urban living.

The quest for clean and safe drinking water is a fundamental global challenge. Robotics plays a role in water treatment and distribution systems, ensuring the quality and availability of this critical resource. Autonomous underwater robots can inspect water infrastructure, such as pipelines and reservoirs, identifying leaks and contamination sources. In disaster-stricken areas, water purification robots can provide emergency access to clean water, addressing a vital humanitarian need.

In addressing global challenges, collaboration and innovation are essential. Robotics researchers and engineers work alongside experts from diverse fields, including environmental science, healthcare, disaster management, and agriculture, to develop tailored robotic solutions. These interdisciplinary efforts yield groundbreaking technologies that have the potential to transform industries and improve the quality of life for people around the world.

As we navigate the complex landscape of global challenges, robotics offers a beacon of hope and innovation. By leveraging the power of automation, artificial intelligence, and human-robot collaboration, we can make significant strides in mitigating climate change, enhancing healthcare, responding to disasters, ensuring food security, and addressing other pressing issues. The role of robotics in addressing global challenges is a testament to human ingenuity and our capacity to create positive change on a global scale.

Conclusion

In the pages of "Mastering Robotics Research: From Enthusiast to Expert," we embarked on a transformative journey through the captivating realm of robotics. This comprehensive book bundle, comprised of four distinct volumes, has been carefully crafted to guide you through the intricate world of robotics research, regardless of your starting point.

In Book 1, "Introduction to Robotics Research: A Beginner's Guide," we laid the foundation for our exploration. We introduced you to the exciting universe of robotics and provided you with a solid understanding of its fundamentals. From the history of robotics to the core concepts and terminologies, we ensured you started your journey with confidence and enthusiasm.

Book 2, "Fundamentals of Robotics Research: Building a Strong Foundation," was your stepping stone towards mastery. We delved deeper into key topics, from kinematics and dynamics to sensors and actuators, equipping you with the essential knowledge required for meaningful research. By the end of this volume, you possessed a robust understanding of the theoretical underpinnings of robotics.

As your expertise grew, we ventured into more intricate terrain in Book 3, "Advanced Techniques in Robotics Research: Becoming a Specialist." Here, we explored state-of-the-art robotics technologies, including computer vision, machine learning, and control systems. This volume honed your skills and offered a glimpse into the exciting possibilities that awaited you as a specialist in the field.

Finally, in Book 4, "Mastering Robotics Research: From Enthusiast to Expert," you reached the pinnacle of your journey. Drawing upon the knowledge acquired in the previous volumes, we dived into complex research areas, encouraging you to innovate, experiment, and create. By the end of this volume, you had transformed into a confident and knowledgeable expert in the realm of robotics research.

Throughout this book bundle, we emphasized the importance of practical application and hands-on experience. We provided real-world examples, case studies, and projects to ensure you could apply your newfound knowledge in meaningful ways. Robotics is not just a theoretical discipline; it's a dynamic field that thrives on innovation and creativity.

As you reflect on your journey from an enthusiast to an expert, remember that the path to mastery is ongoing. Robotics is a field that constantly evolves, driven by curiosity, collaboration, and the desire to push boundaries. The knowledge and skills you've acquired through this book bundle have laid a solid foundation for your future endeavors.

Whether you aspire to conduct groundbreaking research, contribute to technological advancements, or simply explore the endless possibilities of robotics, you now possess the tools to do so. Your journey doesn't end here; it's a launchpad for limitless exploration and discovery.

We hope that "Mastering Robotics Research: From Enthusiast to Expert" has been a source of inspiration and guidance throughout your quest. As you continue to explore the ever-expanding horizons of robotics, we encourage you to share your knowledge, collaborate with fellow enthusiasts and experts, and contribute to the exciting future of this remarkable field.

With the world of robotics at your fingertips, there are no limits to what you can achieve. Embrace the challenges, seize the opportunities, and let your passion for robotics continue to drive you towards becoming a true master in this captivating domain. Your journey has just begun.

About Rob Botwright

Rob Botwright is a seasoned IT professional with a passion for technology and a career spanning over two decades. His journey into the world of information technology began with an insatiable curiosity about computers and a desire to unravel their inner workings. With a relentless drive for knowledge, he has honed his skills and expertise, becoming a respected figure in the IT industry.

Rob's fascination with technology started at a young age when he disassembled his first computer to understand how it operated. This early curiosity led him to pursue a formal education in computer science, where he delved deep into the intricacies of software development, network architecture, and cybersecurity. Throughout his academic journey, Rob consistently demonstrated an exceptional aptitude for problem-solving and innovation.

After completing his formal education, Rob embarked on a professional career that would see him working with some of the most renowned tech companies in the world. He has held various roles in IT, from software engineer to network administrator, and has been instrumental in implementing cutting-edge solutions that have streamlined operations and enhanced security for businesses of all sizes.

Rob's contributions to the IT community extend beyond his work in the corporate sector. He is a prolific writer and has authored numerous articles, blogs, and whitepapers on emerging technologies, cybersecurity best practices, and the ever-evolving landscape of information technology. His ability to distill complex technical concepts into easily understandable insights has earned him a dedicated following of readers eager to stay at the forefront of IT trends.

In addition to his writing, Rob is a sought-after speaker at industry conferences and seminars, where he shares his expertise and experiences with fellow IT professionals. He is known for his engaging and informative presentations, which inspire others to embrace innovation and adapt to the rapidly changing IT landscape.

Beyond the world of technology, Rob is a dedicated mentor who is passionate about nurturing the next generation of IT talent. He believes in the power of education and actively participates in initiatives aimed at bridging the digital divide, ensuring that young minds have access to the tools and knowledge needed to thrive in the digital age.

When he's not immersed in the realm of IT, Rob enjoys exploring the great outdoors, where he finds inspiration and balance. Whether he's hiking through rugged terrain or embarking on a new adventure, he approaches life with the same inquisitiveness and determination that have driven his success in the world of technology.

Rob Botwright's journey through the ever-evolving landscape of information technology is a testament to his unwavering commitment to innovation, education, and the pursuit of excellence. His passion for technology and dedication to sharing his knowledge have made him a respected authority in the field and a source of inspiration for IT professionals around the world.